Opening Africa:

The life of

James Martin

www.OpeningAfrica.com

Opening Africa:

The life of

James Martin

from finding Obama's tribe to founding Nairobi

Philo & M.J. Pullicino

www.OpeningAfrica.com

967.601092

First published in Great Britain in 2008 by
MPI Publishing, Great Britain

Copyright © M.J. Pullicino 2008
All rights reserved

No part of this publication may be reproduced,
stored in a retrieval system or transmitted, in any form
or by any means, electronic, mechanical, photocopying,
recording or otherwise, without the prior
permission of the publisher.
A catalogue record for this title is available
from the British Library.

ISBN 978-0-9544906-2-1

Cover illustration: Emma Pullicino
www.OpeningAfrica.com

M.J. Pullicino has also published:

Process Think: Leading to Change and Innovation (2003)

(www.Process-Think.com)

For my family

Acknowledgements

We would like to thank Dr Andrew Robertson, Dr Richard Manche, Wendy Toole, Laura Pullicino and Kate Pullicino for their kind contributions.

Also Timewell Publications/C.S Nicholls for kind permission to print the photos.

Contents

Map xi

Introduction 1

1. Leaving home 3
2. Two shipwrecks 16
3. Into Zanzibar 23
4. Learning from freed slaves (1880) 29
5. Discovering the Maasailand with Joseph Thomson, FRGS (1883–1884) 42
6. The 'awakening' of Obama's Luo tribe 51
7. The partition of East Africa (1885–1886).................. 71
8. Caravan safaris (1885–1886)...................................... 74
9. The long supply trek to Uganda (1889–1894)............ 80
10. Developing Ravine as a District Officer (1894–1904).. 96
11. Founding a family and a city (1896) 98
12. Benefactor in Entebbe (1904–1914) 107
13. Out of Africa 118

Epilogue 1: The author's discovery of Martin's story 128

Epilogue 2: A token of tolerance 135

Appendix 1: Tribes and languages of East Africa 141

Appendix 2: Journeys on foot in East Africa performed by James Martin 143

Appendix 3: Martin's age 146

Notes.. 147

Bibliography... 148

Author 152

Photos 155

James Martin and his wife Elvira, 1897

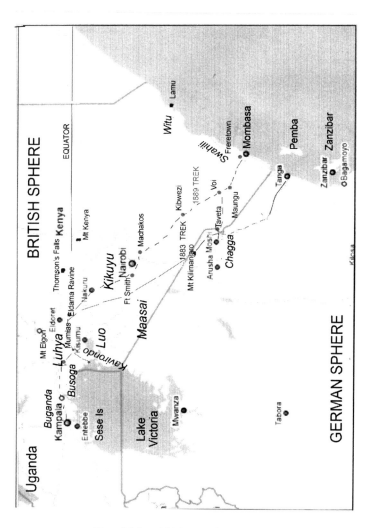

East Africa, 1896: Martin's Treks

Introduction

The Geographical Journal, Vol. LXVI, No.1
July 1925

The Royal Geographical Society, London
Extract from the Obituary of James Martin

> This man stands out as the greatest caravan leader of the day when the lives of Lugard and his staff in Uganda were absolutely dependent on the periodic dispatch of stores transported by large bodies of native carriers from Mombasa – a trifling walk of 800 miles.
>
> Martin was a Maltese by birth and a sailor by trade. A man of no education but possessing great natural gifts and no mean ability, his connection with the East Coast of Africa commenced about 1880 or perhaps earlier. He was a hand on a sailing vessel which was wrecked in the Red Sea …

With the coming of dawn the storm seemed to be abating. The gusts of wind, low over the surface of the sea, were less strong, and the waves had lost something of their whiplash. This part of the Red Sea was prone not to high daunting waves but to shorter, more punchy ones every bit as dangerous. The sky was still leaden with low heavy grey cloud and did not promise any quick relief.

Opening Africa

The shipwrecked sailor clung desperately to his spar and for the umpteenth time he checked, in an almost automatic movement, the end-frayed rope which had providently been entangled with the spar when he grabbed it in the water and which he had wound around his chest and under his armpits as his first reaction to seek safety from drowning. Indeed, without that lifeline he would long have been parted from his improvised raft.

His first contact with the dark black sea had been a shock, but it had also triggered his swimming reflexes and he was quickly able to break through to the surface and gasp fresh air and breathe again. He was a good swimmer, having learnt to swim at a very early age, but even the best swimmer would not have survived for long in that turbulent sea without some means of support.

He had been in the water for probably four or five hours in pitch darkness since that frightening moment when he was washed off the deck by a gigantic rogue wave that struck as the ship was heeled on her beam ends. Part of the mainmast had snapped and various pieces of deck equipment had been wrenched off and cast into the sea before the vessel itself disappeared in the darkness ahead. He was not sure whether she had righted herself and sailed on or whether she had foundered. There was no way of telling. But then, just as he was cursing his bad luck, he bumped against a larger spar that had broken away from the mast with part of the rigging.

Now he was miserable and cold, with aching limbs. He felt as if he had been in the water for days, and as his strength weakened so did his recollection of time and events. His mind kept wandering back to the past. Why did he not stay at home? Why was he here?

1

Leaving home

He pushed his head over the thick yellow limestone bastion walls and looked down into blue expanse of the Grand Harbour. He was at the 'Ix-Xatt tal Belt' not far from the old customs house. He had come to see the latest ships that had arrived, as well as the British Mediterranean fleet. He was always fascinated by the big ships anchored here and knew what most of them were. It was magic to him – all those flags and sails caught his imagination. The harbour had never been so full with every kind of ship. That was the moment when he said to himself that he would definitely go to sea, as soon as he could, probably in one of those magnificent three-masted ships. Although he was only fifteen, he felt he was ready for it.

    His father was a sailor and had left on one of his many trips from that same place. Family goodbyes were never easy, and the last one had seemed even harder. He expected his father back any day now and wished he could be back sooner. He looked at the ships in anticipation. He got most of his information about them from his father, who always answered his questions. It seemed that he never had enough time to ask all he wanted to ask before his father was off on his next trip.

    The Suez Canal had opened just three years previously in 1869, and now that many more ships were passing through Malta on their way to the Far East it had, in a very short time, become a leading world harbour. He could see that there were mainly sailing ships in the harbour, some like the great clippers that occasionally (although rarely) came in; but he noticed also that steamships were becoming more commonplace, and the most impressive of these were the latest iron steamships. There were some that used sails but were steamships at the same time. He often wondered which propulsion system they used most and why – questions he was keeping for his father. For the second time he saw the new India mail steamship, which had three masts and one funnel. It had all the latest equipment: he had been told about the details by an English navy officer only the week before. These ships had been specially built for quick passage through the Suez Canal.

Not all the ships were bound for trade with India and the Far East. It was not long ago that transport ships carrying the wounded from the Crimean War had anchored here. Those who were too sick to travel back to Britain had recuperated in the Malta hospitals before being shipped home.

Warships, trading ships and passenger ships sailed in and out, but the largest ships flew the Union Jack, as Britain had officially administered Malta for nearly fifty-five years now. In spite of this, English was still not as widely spoken as Italian, and when it was it was by the upper classes. His father, being of Italian descent, often spoke Italian to James – as well as Maltese, of course – so the boy had learnt to get by in it.

James was fortunate in that the sail factory where he worked was run by an Englishman, and as all the technical sail jargon was in English he had managed to pick some up. He had been working there for nearly

Leaving Home

seven years and had mastered sail-making with all its stitches and knots. His precise hands were good at the tight stitching and he knew that he had security there, but he was starting to find it boring.

*****

His mother always seemed to be worried. Often it was about water. The rain would collect on their roof and run into the house well, but in the late summer, when there hadn't been much rain, their well ran out, which made things difficult. And the water they did have was not always the cleanest. However, the Knights had, around 1620, built a long aqueduct that supplied Valletta with water from the hills near Rabat. It was in fact built by Grand Master Wignacourt who also paid for the sixteen watch towers around the island. Luckily the aqueduct passed not far from his village, and James knew a way to get his family water from it so they never went without.

    She also complained about the price of bread. The baker said it was because of the wheat tax, which kept going up. She seemed to always be busy, either looking after younger children or suffering from the effects of pregnancy; a couple of her children died at birth. She thus had little time for James, and this enabled him to more or less do as he pleased. He had, from an early age, learned to cope without his mother or father, and he constantly had something going on with one of his friends.

    The farthing school fees at the local Catholic school were too much for his mother to afford, and in fact very few of his friends in the village of il Marsa (which means 'port' in Arabic) went to school – probably less

Opening Africa

then 5 per cent.

'What do you want to go to school for?' his mother used to say. 'It's a trade you need to learn, so you'll be able to earn your living.'

They lived in a little flat-roofed limestone house, down a narrow side road. It was just a few rooms on the ground floor, with a courtyard, and neighbours above. The local priest often came to the house and talked to James's mother about her children's education:

'If not the girls, send your boys to the local school, or at least your most gifted one, James,' he would try to persuade her.

'We can hardly afford bread, let alone schooling,' she told him. 'If we send one boy to school we have to send them all.'

James felt deep inside that he would like to learn to read and write, and that it was an injustice that he was unable to do so. Although he had the basics for making a living, and didn't really go hungry, he felt that he needed more. He longed for the sort of recognition and self-respect that he couldn't achieve with what he was doing now.

His mother didn't speak any Italian; she spoke only Maltese to the children. At this time the local language was described by the English as 'a sort of Arabic dialect'. The Maltese believed it to be their own unique language, and few knew that it was in fact a legacy from the Arabs who had lived on the island nearly a thousand years earlier.

James had gone to the local 'skola tan-nuna' (nursery school) up to the age of seven, where he was taught children's folktales, nursery rhymes and prayers, all mixed into one. That was just about all his 'school' education, apart from the local priest's religious classes on Sundays, which he didn't like and didn't always attend.

Although formal religion wasn't for him, he did take in the main essence of its values and beliefs. God was compassionate and loved. The family and the marriage unit were important for happiness. He had to avoid temptations that led to sin (excessive drinking was one of them). All men were equal in God's eyes. He had repeated these principles by rote many times and in different forms at Sunday school. He sort of understood them, but he wasn't really sure whether they had any relevance to him and his life. He bore no grudges, but he understood that religion was religion and life was life. He certainly didn't see much evidence in reality of the last of these beliefs, that all men were equal.

It didn't seem to matter for James that he couldn't read; less then 8 per cent of the population could read Italian, and less then 4 per cent could read English. Of course, many more could speak these languages. Maltese was spoken by nearly everyone but was read and written by very few. However, James could get by speaking all three languages, and that ability of his was quite special.

As he couldn't read books, he learnt how to read faces. He was quick-witted and knew how to bring a smile to someone's face. His skill with languages was the key. He knew how to communicate and thus how to find things out. Just last week when he had wanted to know about the new India mail steamship, he had gone up to the hill known by the Navy as 'Nix Mangiare Hill' (nothing to eat) near the Fish Market. It was where boys and young men went looking for work, or begging for food or pennies, near the approach to the main Victoria Gates of the city of Valletta. There he spoke to an English navy officer he saw by the shore. (He was not as shy as his friends were in approaching an Englishman.) If he'd thought he was likely to get brushed off, he was wrong.

'Good morning, sir,' he said. 'Fine morning.'

'Hello, son. Yes, it is nice and sunny today,' replied the officer.

Pointing at the new ship, the boy remarked, 'That's a strange ship over there. If that funnel was any bigger they could use it as a sail!'

At that the navy officer and James burst into laughter simultaneously. By coincidence the officer was pushing to get more technology into the navy's ships, and that funnel was ridiculously large, so the joke struck a chord. They fell into conversation, and the officer explained in detail about the steam turbines that were the 'latest thing' fitted on ships, and how they drove dynamos, and how these in turn made electricity and light for the ship. James went home happy knowing that he had got so much information, and that most of it couldn't even be found in books.

This was his usual way of finding things out. He would look at a person's face, talk to that person in his or her own language and make a joke: that opened the person up and allowed James to ask all the questions he could want. A good way to make a simple joke was to find two parallel meanings or functions of a word and make a play on them. It always worked. He used to remember the information he was given together with an image of the person's face, as a sort of reference. He found he could remember almost anything that way. 'Who needs to read to find out about something?' he thought to himself. He did it his way, and would in time perfect and stubbornly stick to this 'information retrieval system'.

Although schooling had not given him rules and structures to learn things by, he found that at different stages of his life he occupied himself with learning different topics. In his quest to speak languages he found that there were certain areas about which it was important to be able to talk: in addition to exchanging greetings, you needed to be able to discuss food, travel, family and trade (which included money and counting). He always wondered how they learnt these things in school and was more and more determined to perfect his own way.

He spent much of his time, when he was not working at the sail factory, playing with his friends. He noticed that his skin was fairer then some of theirs, probably because of his Italian or Sicilian ancestry (although his family had been in Malta for generations). One of his best friends had a rather darker complexion, and in the height of summer, when they had been in the sun and done a lot of swimming, his skin turned almost black, perhaps reflecting his North African origins. It wasn't an issue. They were all Maltese. People James knew had surnames like Hassan, Buhagiar, La Rosa, Portughese, Portelli, Stricktland, Borg: these names were found in places from North Africa to northern Europe, and reflected the time when Malta was ruled by the Knights Hospitaller.

It had occurred to James that living on or near 'the other side' (the south) of Malta was a disadvantage. The people were poorer, and it seemed that it would be impossible for him, as a member of the 'lower classes', to move from his low-paid sail-making job to be one of those well-to-do 'signuri' who went around in the horse-drawn kallezzis, or who could afford to travel in omnibuses (also horse-drawn) and live in large houses in Sliema. He liked horses and for that reason enjoyed being in il Marsa, which was something of a centre for stables. He had spent many hours helping to clean them out and earning a bit of cash. He wondered if he would ever be able to have his own horse. In his present position, he did not see how this could happen.

*****

As he got older he became more interested in the social and political aspects of the British being in Malta, and he was sensitive to how the Maltese treated one another in the different social classes. That it was just about the highest point of prosperity of the Victorian age and the British Empire had a bearing on what happened in Malta. Victoria had been on the throne for almost thirty-five years. The way of life of old Georgian Britain had shifted and the country was now leading the world in technology, commerce and prestige. It was also the greatest military power. The British Navy controlled the oceans and had a dedicated Mediterranean fleet with its head quarters in Malta. While other competing empires – the Russian and the Ottoman – were still in an era of feudal serfdom and corrupt practices were rife, the British Empire's goal of carrying on honest and benevolent government was as important to it as its commercial ambitions.

With Italy to the north now united under King Victor Emmanuel II – the culmination of events that had started with Garibaldi's taking of the neighbouring island, Sicily, ten years previously – James could recognize more ships with Italian flags in the harbour. His father had often spoken of '*il Risorgimento*', or the reunion of Italy, as being a good thing, but Italy was still in turmoil and there was no real prosperity for the lower classes.

All in all, it seemed, Malta was lucky that it had 'asked' the British to administer the island, and James had some understanding of how the British governed and helped the Maltese. They built up the infrastructure and generally, by working together with the people, made Malta a better place to live. They provided many jobs, and not only in the dockyards. The British of course had the use of the Grand Harbour for their ships and imperial ambitions. They were in turn impressed by the buildings and institutions in Malta, such as the judiciary, whose official language was Italian, and the university, which they respected and fostered. This knowledge of British administration was to serve well later on in life.

Working-class families growing up in Malta at that time, although having to live close to the breadline, were offered opportunities of work in the port and dockyards and in the industries that serviced the navy (such as sail-making), or they could work in the fields. But there were some, even in his own village, who complained about the poverty of the working classes and their inability to change anything. In reality, they said, the British supported the upper classes, and the rest were the losers: 'Once poor always poor.' It was the same with education – a lot was being talked about and done to improve the education system, but there was nothing for the lower classes. It had been much the same during the times of the 'Knights of Malta' before the British came, when the Maltese working classes had been the foot soldiers. (Indeed slaves had existed in Malta up to 1798, when the Knights were overthrown by Napoleon, who then freed the slaves and abolished slavery on the Island). There were groups in Malta that were secretly calling for self-rule.

It was said that there were two kinds of Maltese: those that left and those that stayed. At this time nearly 15 per cent of the population had emigrated over the previous half-century. Some, like James, were 'pulled' by the opportunities that the wider world offered, but most were 'pushed' by the poverty and conditions in their limited homeland, which was often unable to provide for its 140,000 inhabitants.

There were many things that indicated to James that he should go, but also others that told him he should stay at home. Among the positives, he remembered with joy the traditions and festivities of the island that gave him a sense of belonging. He worried that if he left he would lose these. Every Saturday afternoon and evening many people from the villages would walk up and down the main street in Valletta, and once the gas lamps were on it was possible to stay out until late in the flickering light. That's where everyone got the news of what was happening. It was the place to meet girls from the neighbouring villages as well. Everyone just walked up and down and had a good time. The fastest and latest horse carriages were to be seen. He had once taken the horse-drawn omnibus home with about fifteen others – it used to run between the villages and Valletta – but walking, which he enjoyed, was the way he usually got about.

The feast day that occurred every year was also a great time in his village. Everyone would be out on the decorated streets to see the processions and the local band, and there was singing and folk dancing until late

Leaving Home

into the night. Everyone would dress up in their finest clothes: the gentlemen in their dark suits, the menfolk in their best trousers and a white shirt, and the ladies in their brightly coloured Malta cotton garments. The other big annual event was the carnival in Valletta, when everyone went to watch the colourful floats. This time people dressed up in fancy costumes and there were many different bands to listen to. It was all great fun, and at times like these the island seemed like one big happy family.

The bright colours of the floats contrasted with the 'faldettas' that some of the women wore: full length black gowns with flat cardboard headpieces that were supposed to catch the wind and cool their faces down. His older cousin had told him to avoid these women because they were spinsters, and the superstition was that they could put a spell on you. He had always crossed over to the other side of the road when he passed one of them, and would secretly point his covered first and fourth finger at her and say a little rhyme to ward off the evil eye. If all this had happened while he was passing a cemetery, it had been even more frightening. It seemed stupid to him now that he was growing up, and he just didn't believe it anymore, but it had made him frightened at the time. It was good to see the faldetta ladies joining in the festivities on these occasions – it seemed they were human after all.

These, then, were the good times. Other thoughts brought a heavy feeling to his heart. He had seen the increase in the number of people around the harbour cities of Senglea. People moved out because the area was too crowded and the drainage systems didn't work well; there wasn't enough water to waste on flushing the drains, especially in summer when it was the hottest. Sometimes the whole place stank to high heaven; il Marsa, being low-lying, was particularly bad. Complaints to the administration never had any effect.

This problem with the drains was the cause of several bouts of disease that killed large numbers of people. The smallpox outbreak in 1830 was still talked about, as was the cholera of 1837 when over 4,000 people (3 per cent of the whole population) had died. The last cholera outbreak had been not so long ago, in 1867, and had killed almost 500 people. Everyone had been warned about the symptoms: a sudden onset of watery diarrhoea, stomach cramps, nausea and vomiting. There was only a 50 per cent chance of survival if you got the disease. James also knew the symptoms of smallpox: high fever, headaches and vomiting followed by a face rash. The authorities insisted on being informed if anything like this occurred. Then it wasn't just the sick who were affected. The quarantine measures that were brought in stopped all the trade and put people out of work. His own sail-making company, it was said (although this was before his time), had had to lay people off when there was an outbreak of disease as fewer ships had come into port to get new sails and the business had gone into a decline.

The Crimean War (1853–6) had led to a massive increase in British expenditure in Malta, resulting in jobs for nearly everyone, but that was now over and in the last fifteen years about 15,000 people had left to seek a living elsewhere: that was roughly 12 per cent of the population. This migration was a continuing trend, and by the end of the nineteenth century nearly 20 per cent of the Maltese population would be living around the shores of the Mediterranean. Without realizing it James

Leaving Home

had understood this trend, and felt that he too would be better off away from the island. Such decisions are much easier to make as a fifteen-year-old with no commitments or ties.

Many people left Malta because it couldn't provide them with enough food. James Martin left because it didn't provide him with enough opportunities and challenges to fulfil his dreams. He would carry with him his Maltese cultural DNA, which would be severely tested in the years ahead.

2

Two shipwrecks

James Martin was now a deck hand aboard a three-mast barque that had sailed from Liverpool four weeks previously, bound for the Indian port of Bombay. Although the summer months were long past, the weather had been kind in the Bay of Biscay and the wind in the Mediterranean was favourable. There had been a short stop at Gibraltar, and another at Malta for water and provisions. This last had been particularly welcome for Martin as he had looked forward to returning to the island. He was thrilled as his vessel entered the Grand Harbour to be welcomed by the enveloping embrace of the towering bastions and battlements of Valletta on one side and the Three Cities on the other.

  The 'three cities' consist of Birgu (Vittoriosa), which was the original city inhabited by the Knights of Malta when they came in 1530. It was granted the title of Citta Vittorosa in 1565 after the defeat of the Ottoman Turks in the Great Siege of Malta. The next, Bormla (Cospicua), has the designation of Città Cospicua. It was granted this title by Grand Master Marc'Antonio Zondadari in 1722. The final one is Isla (Senglea). These three cities formed the main urban and political centre of Malta before Valletta was built.

  On seeing the Grand Harbour, childhood memories came back to him. As a boy he had known the many inlets and creeks around the island's coastline and had come to love the sea automatically; fishing, swimming and diving for sea urchins were part of his life. He remembered how he had spent much of his time on the rocky shores of the two main harbours around Valletta, watching with awe as the large sailing vessels manoeuvred gracefully within those confined waters, backing sails, hauling them down, adapting sail area to the wind and to the particular task in hand. He studied the interplay of anchor and sail necessary for the vessel to reach a desired anchorage and the precautionary use of a whaler or two when the vessel was ungainly and did not readily respond to the rudder.

He had been away from his old sail-making job for what seemed a long time now, but he had often been told that the quality of the Maltese cloth was such that countless sailing vessels came from far and wide specially to renew or to replace their sails at Valetta. The sailcloth industry had grown over years, as had the production of the prized variety of cotton that adapted itself so well to the island's poor soil. Scores of workers in the cotton fields and in the weaving shops of Zebbug and Qormi made a living from the industry, although cheaper cotton from Egypt was increasingly being imported, to the detriment of the local cotton farmers.

At the age of sixteen James Martin had been taken on by the skipper of a merchantman that was calling at Malta, and after a tearful farewell to his family, who showered him with warm clothing, gifts and good advice, he embarked on his first voyage. His talents were soon recognized, and as well as carrying out all the dreary chores assigned to a young newcomer to a ship, he automatically made friends with the ship's tailor, who also cared for the sails, the ropes and the rigging.

Opening Africa

He soon became adept at handling sails, learning their various names and getting to know how to repair, sew and replace every bit of needy canvas, as was customary on a sailing vessel that had a limited crew and trades. All this was to serve him well in his future life.

He had no difficulty changing ship when he arrived in Liverpool at the end of his first engagement. He chose a ship bound for the United States, and moved on to other ships so that he could fulfil his desire to sail the seas and see the world. He grew in years and strength, and developed into a hardy and confident seaman – a friendly and popular figure on board and on the wharves, and in seaman's haunts in the larger ports.

At birth he had been christened Antonio, after his father. His real surname was Martini, not a common name in Malta, and it indicated his Italian or Sicilian ancestry. The Martinis had, however, been in Malta for several generations. As soon as he went afloat, he became 'Jimmy' to his friends. Martin didn't mind – in fact, it was the trend in Malta at that time to be called by an English name instead of a previous Italian one. When Jimmy Martin was shipwrecked in the Red Sea he was twenty-three years old.

The passage of the barque through the Suez Canal was slow and laborious, involving much towing from the shore and hauling by the ship's boats. The canal, although shortening the journey to the east, was not very practical for sailing vessels. Out of Port Suez, the barque spread her wings with the zest of a caged bird set free and revelled in the strong winds even as they rose to near gale force. The skipper caught the exhilarating infection of speed, and used all the sail he could to get every advantage from the favourable run. Alas, he failed

Two Shipwrecks

to temper his enthusiasm as the weather deteriorated and refused to heed the warnings spelt out by the banshee-shrieking of the rigging, by the protest of the creaking timbers, and by the repeated advice of old, experienced hands.

The skipper held on, just as a gambler holds on to a winning streak. He was foolish, especially as the light was poor. Then suddenly it happened. The mainmast, heavily over-canvassed, could take no more. Part of the rigging snapped just as the vessel was heeling hard to port with gunwales awash, and the top fifteen feet of the mast snapped off with a loud crack, flying overboard and dragging with it rigging, spars and shredded sail cloth. As the ship, relieved of her topsail, swung violently over on to the starboard beam, right into the trough of a rogue wave, the sea swept viciously across the deck. There was no hope for Martin the sail-maker, who at the time was attempting to reach the shelter of the wheelhouse. Caught unsecured in the open, he was swept into the turbulent waters of the Red Sea in full storm.

*****

HMS *London* had just passed through the Suez Canal on her way to Zanzibar, making far better time than Martin's sailing ship. She had called at Malta for coaling and her next stop would be Aden. The captain had made the journey before and knew how vessels of various sizes and nationalities converged on the eastern end of the canal. As a precaution he doubled the lookout watch and found that it gave him greater peace of mind, for many of the vessels he encountered were under sail and having a difficult time in the high wind and heavy sea.

Opening Africa

It was one of the hands on extra watch who spotted and drew attention to the bits of wreckage and other flotsam in the path of Her Majesty's ship. Then a spar with a man lashed to it drifted close by the starboard quarter and was miraculously noticed by the lookout. The speed with which the wreckage and the ship converged and parted was hardly sufficient for the lookout to ascertain whether the human being he had seen was a corpse or was still alive, and it was a fortuitous decision on the part of the captain, who happened to be near the wheelhouse, to give the order to slow down and go about to enable the crew to have a closer look at the wreckage.

Martin was only just aware that he was being picked up. He could not, however, fathom out where he was or why his aching body responded so willingly to the comfort and warmth of the mattress and blankets in the ship's sickbay, the like of which he had not experienced in the cramped sleeping quarters and hammocks of the ships in which he had sailed. It was a whole day before his befogged mind could comprehend his extraordinary luck. He had not broken any bones and he was on his feet.

By the time HMS *London* reached the port of Aden he was fit enough to be put ashore, and bidding farewell to his saviours and thanking them for the clothes they had given him, he went to a small and cheap seaman's hostel run by a Yemeni Arab. Inquiries revealed no news of Martin's stricken sailing vessel, and he assumed that she had foundered. Within a week he had signed up on a merchant steamer bound for Bombay, and after a short stay in that city, to which he took a dislike, he found an opportunity to sign up on a smart

Two Shipwrecks

American steamer bound for Zanzibar. Zanzibar had not previously entered his mind – it was an out-of-the-way place, its name conjuring up strange people and curious customs – but this move fitted in with his plan to serve on different kinds of vessels and travel the world. He might also be able to link up again with the blue-jackets of HMS *London,* who had saved him in the Red Sea.

Martin had not reckoned on the American skipper of the steamer being a drunkard and in every way an impossible person. He was burly, tough, and disciplinarian to the point of cruelty. The ship was sailing in ballast, but there was certainly no shortage of liquor on board as the captain was never sober throughout the voyage. He bawled out his orders and cursed and swore without provocation at one and all. Even the ship's cat was a target of his fury: no wonder all the crew were cowed and feared him.

As the ship crossed the Indian Ocean he muttered unintelligibly to himself and seemed bent on some mischief, busy with his charts and his sextant. On the day before they were to reach land the captain's drinking increased. He set the vessel's course on a straight line pointed at the eastern coast of Zanzibar, made the helmsmen – among them James Martin – swear that none of them would alter the course of the ship in the slightest degree on pain of being severely punished, and retired to his cabin hugging a couple of bottles. He locked himself in, with orders that he should not under any circumstances be disturbed. It was not clear if his plan was to wreck the boat and claim on insurance.

When the danger arose and the presence of the skipper was needed, he was completely lost to the world. The coral reef, the shoals, the sandbanks all loomed suddenly out of the darkness. The charts were locked in with the captain, and the vessel charged on to the inevitable fatal destination that the captain had planned for her. Martin was at the wheel. He did his utmost to slow down and save the ship, but with little effect. Luckily it was low tide, and the ship ran aground on a sandy beach with a shattering bump. She was held fast at the prow and miraculously was not badly damaged. She stayed more or less upright; the problem was to get her afloat again.

This disaster had been witnessed by the crew of a Royal Navy cutter that was hiding up a narrow creek near by, screened by mangrove trees. The cutter, one of two in Zanzibar waters, was on anti-slave patrol, and had been on the lookout for the slave-laden dhows that continued the now illegal trade that the British government had resolved to suppress. The crew of the stricken American vessel put out stern anchors, and as the tide rose warped the ship with the help of the British blue-jackets. Slowly and with much effort the ship moved; the rising tide, twelve feet of it, was a powerful and irresistible lifting force. Before long the vessel could go astern and be taken for damage inspection to Zanzibar Town, about fifty miles away on the western side of the island.

3

Into Zanzibar

The 1870s and 1880s brought about very important developments in east and central Africa. The main causes were the opening in 1869 of de Lesseps's canal joining the Mediterranean to the Red Sea – the Suez Canal – and the ever increasing number of 'steamers', ships that relied on steam-driven propulsion instead of, or in addition to, sail. These novel facilities fired the imagination of shippers and skippers to open up new sea routes, and little-frequented ports like Suez and Aden grew in importance as stations for coaling and the replenishment of stores.

Up until then the inhospitable and unfrequented coastal waters of East Africa, from Cape Guardafui to Mozambique, had held little attraction, except perhaps to the dhows that since time immemorial had sailed from the Persian Gulf to Zanzibar and back, relying on the prevailing winds to carry their miserable cargoes of slaves to Arabia, to India and to Persia. The early explorers of East and Central Africa – Livingstone, Speke, Burton, Grant and Baker – either travelled by sea via the west coast, round the Cape of Good Hope, or overland up the River Nile from Egypt and Khartoum. Now that the new route via Suez made the journey so much shorter, the importance of Zanzibar grew rapidly as a stepping-off port for an increasing number of European missionaries and explorers, mainly British and German, who were anxious to carry the Christian faith to pagans and to draw maps of unknown areas of the world, which in Africa, since Roman times, had begun with the undefined boundary of the legendary Mountains of the Moon.

Great Britain, the leading maritime and naval power at the time, built up her influence in Zanzibar and the surrounding seas sufficiently to be able in 1873 to persuade the Sultan of Zanzibar, Seyyid Barghash bin Said, to prohibit if not slavery itself then the carriage of slaves by sea. This ban was enforced by a small fleet of Royal Navy cutters and whalers that patrolled the Indian Ocean off the east coast of Africa, particularly around Zanzibar and Pemba. The Royal Navy sailors, under the command of young officers, had many a skirmish with slave traders on the high seas, but it was difficult for them to free the captives since the slavers disposed of their human cargo by jettisoning it into the sea whenever a naval patrol was sighted.

The crews of the naval boats were based on Zanzibar, where on nearby Grave Island memorial stones to this day bear witness to the courage of those who died in performance of their duty. Stores and supplies, arms and ammunition, and replacement crews would be provided by a Royal Navy depot ship anchored in Zanzibar Harbour. HMS *London*, which snatched James Martin from certain death by drowning in the Red Sea in 1879, fulfilled such a mission for seventy years from 1874.

In the second half of the nineteenth century, Zanzibar was in many ways anything but attractive. A visitor arriving there described the scene of the foreshore as 'cluttered with rubbish and with pariah dogs scavenging among the corpses of slaves who had dropped dead there or had been callously discarded by their owners'. The Stone Town was a cramped collection of narrow streets, layers of wooden balconies, dirty walls, overhanging rusty tin roofs, and closed doors everywhere; some of the latter, elaborately decorated with brass studs, hinted at the wealth that lay behind them.

The air had a distinct smell that was both attractive and repulsive. The pleasant scent of tropical fruit, cloves, nutmeg, cardamom and other spices, and sometimes the whiff of fresh tropical rain, was all too often overpowered by the decaying stink of sweat and sewage, depending on the direction of the wind. It was an odour that stayed in the mind long after one had left the island.

Zanzibar was the gateway to the African mainland, which beckoned with dreams of spoils and riches but also with the terror of hostile inhabitants, incurable diseases and inaccessible places. It was a land of life and death: in the first half of the nineteenth century, more than half the Europeans who ventured in died of diseases such as malaria. Of those that returned from the mainland, some had gained instant fortunes in ivory, gold and black slaves. The myth of the 'Heart of Darkness', as it was called, was kept alive.

Yet there were Europeans who fell under the spell of Zanzibar and, once there, did not want to leave. One such man was Sir John Kirk, who in 1858, aged twenty-six, joined the great missionary explorer David Livingstone as his doctor and botanist. When Kirk returned to the United Kingdom seven years later he was a tired and sick man, but as soon as he had recovered he sailed back to Zanzibar to work as a doctor and Vice-Consul, there to remain for over twenty years. He retired as British Agent and British Consul General, and a close and trusted friend of the Sultan, after having firmly established British predominance over the Sultanate.

Another Englishman to be caught by the island's spell was a young naval officer who first saw Zanzibar in 1874, when he was based there in connection with anti-slavery patrols. He was Lieutenant Lloyd William Mathews. Within three years he was inexorably drawn back to the place, and volunteered to help the Sultan in training his military forces, without giving up his anti-slavery patrols. The young officer built up the Zanzibar army from 500 men to 1,300, and was then seconded to the service of the Sultan and subsequently made Commander of the Sultan's Forces. Five years later he was a Brigadier General and the Sultan's Chief Minister.

Along with Sir John Kirk he pursued a relentless policy aimed at the suppression of slavery in East Africa. He led armed expeditions to the mainland to arrest Arab slave traders, to set up military posts under the Sultan's flag, and to sign treaties with various local African chiefs whereby they agreed to recognize the overlordship of the Sultan. Mathews's influence in Zanzibar and on the mainland of East Africa was said to exceed that of the Sultan himself.

Among Mathews's protégés and colleagues was James Martin, whom he had met in Zanzibar soon after the grounding of the American ship. Martin himself was to become deeply affected by the spell of Africa and would never again return either to his seafaring career or to his native island in the Mediterranean. The challenge of adventure would bind him inescapably to Africa.

Into Zanzibar

Zanzibar seemed so different to him. He saw Arab tradesmen dressed in their imposing headgear and gowns, some with ornate curved daggers attached to their belts. The women wore long black gowns that covered their heads (called 'bui bui') – they looked a little the faldettas worn in Malta but were without the flatter headpieces. He found that he could pick out Arabic words and understand the gist of a conversation as the words sounded like his native Maltese; indeed, he could count in Arabic. He could, by changing the accent of his Maltese, make himself understood by Arabs and carry on a basic conversation, which made the place seem less like a foreign land to him.

Martin noticed that he could even understand some of the Swahili that was spoken by the locals. The native Swahili greeting 'jumbo, habari yako' included the word 'habari', which sounded like the Maltese 'ahbariet', meaning news. It meant 'hello, how is your news?' He also quickly found that the Swahili numbers between six and nine were very similar to his own, and indeed to Arabic numbers as well.

He worked out that if he wanted to discover a word in Swahili, he just had to ask an Arab who spoke Swahili, which he could do using his Arabic/Maltese. Thus by using his Maltese he could get into Arabic, and from Arabic he moved into Swahili. He had the ability to remember the different words immediately and to use them himself. This was a revelation to him, and he found mastering different languages an exciting challenge.

*****

Opening Africa

With their background of lush green coconut trees and surrounded by the bright blue sea, the many white sandy inlets, with their fishing boats anchored out or pulled up onto the shore, made an inviting picture. The long narrow boats or 'ingalawas' – dug out from tree trunks, and with a raised bow – had two outriggers. Their single sail hoisted on a gaff rig that caught Martin's eye and roused his maritime curiosity.

But the positive impressions of his first days of being in Zanzibar were soured by other realities. Walking up one of the narrow streets of Stone Town he saw a sight that sickened him to the core. A black boy, probably not more then ten years old, was carrying a heavy log on his head and the log was chained to his ankles. Martin was thus confronted face to face with slavery and found it difficult to understand. The sad, wide-open eyes of the boy were imprinted on his soul and he felt utter despair. The fact was that behind the closed ornate doors and barred windows of Zanzibar, the majority of the black population were slaves. Indian traders owned about 8,000 and the Sultan 4,000 for his clove plantations; other Arab and black citizens owned between 500 and 2,000 slaves.[1]

4
Learning from freed slaves (1880)

The Europeans in Zanzibar numbered only a few dozen: the small staffs of the Consulates of Great Britain, Germany and France, usually limited to one or two expatriates, and the few missionaries at the Church Missionary Society and Roman Catholic missions. When Martin arrived in Zanzibar Stone Town he contacted the British Consul, Sir John Kirk, and discussed the shipwreck with him. There was nothing that Kirk could do about it. The Sultan's administration – what there was of it – would not be in a position to take any action either. It was therefore up to the American skipper of the vessel to report the circumstances to the United States Consulate, but there is no record of any action having been taken as a result.

Martin was billeted with the British sailors until arrangements could be made for his future. He disliked all forms of offensive action or fighting and would not join in the Navy's task; anyway, he could not be absorbed into the Royal Navy, or receive any wages from that source. There was no regular calling of steamers at the island, and it could be a month or two before any arrived. The Church Missionary Society came to his rescue. There was no work for him in its Zanzibar Mission, but the newly established CMS centre in Mombasa, called Freretown, was sorely in need of help to cope with the influx of Africans from various parts of East Africa who sought refuge there after escaping or being liberated from the Arab-led slave caravans. The town was named after Sir Bartle Frere, a former Governor of Bombay, who had helped push through the Anglo-Zanzibar treaty that abolished slave trading. Lieutenant Lloyd Mathews was just about to leave Zanzibar for Mombasa with a small company of local troops to establish new posts on the mainland. He offered Martin a passage, and on the journey the two got to know each other and struck up a friendship that was to last as long as both remained in East Africa.

Freretown was on the western or land end of Mombasa, and beyond it lay the African bush from which the tired and stricken slaves arrived. CMS not only gave refuge to those it welcomed, but sought to rehabilitate them by teaching them the rudiments of some trade or other. Martin was given the task of general supervisor, and was able to make use of all he knew about sail-making, tailoring, carpentering and the other odd jobs that he had handled in the ships in which he had sailed in.

He also learnt how to excavate wells, build huts and set up a camp. The positioning of a camp away from flood danger and away from swamp wetland was the basis. The camp had, however, to be close to a water source such as a river, spring or well. He learnt to allow for a slope and trenches to drain the camp in the heavy rains. The logistics of getting wooden poles and materials for the huts were part of the preparation. A hard wood (usually mvuli) was picked for the main columns of the hut, which were dug deep into the red soil to act as an anchor and stop the hut from being blown down in the high winds. Freshly cut wattle branches, stripped of their leaves, were then woven between the main posts. Strips

Learning from freed slaves

of coconut palm leaf were used to tie everything together. The roof trusses, positioned to form an apex, were also of a hard wood and were tied, again with coconut palm leaf strips, to a kind of ring beam. Next the pitched roof, added to the skeleton structure, was made by hanging coconut leaves ('makuti') with the 'V' facing downwards to let the rain run off. Several layers were tied and built up. Next the wattle walls were infilled with wet mud – just soil mixed with water. This made the walls, once the sun had dried them out, as hard as stone. The final task was to make drains so that the water running off the roofs would be collected and would run away from and not into the huts. The roofs had an overhang of two feet or more, providing shade for those sitting outside the hut.

Latrines were separate and were dug a distance from the huts. The whole camp or 'boma' was protected from animals by an outer circular hedge of cut thorn bushes around the huts. Sometimes a large ditch was dug in front of the thorn bushes for protection and drainage purposes. There was a gap in the thorn bushes for the entrance that was sealed by means of a gate. This was a building pattern Martin would repeat many times in the future and would become expert at. As some of the freed slaves were women and children, suitable separate accommodation had to be built for them. He soon found a certain satisfaction in helping and caring for the unfortunate inmates who had no home or family.

Martin had arrived in Zanzibar with little prior knowledge of the place and its history, and the short stay he had there allowed insufficient time for him to be inducted into the social mindset of the Europeans. He thus arrived at Freretown with an untainted, or perhaps naïve, view of Africans and in particular of slaves. He didn't have the common attitude of white supremacy over black. His communication with them could thus be without the prejudice, and he was free of the prevailing view that slaves were 'wild savages'. Added to this, his Maltese background determined that he approached people with warmth and respect.

Although he never learned to read or write, at Freretown he developed the knowledge of spoken Kiswahili and other African languages and dialects that was to prove invaluable to him in the various tasks and appointments he later held in East Africa. His natural talent and the few Swahili words he had learnt in Zanzibar enabled him to communicate with his Swahili helpers.

Martin realized that the different tribes that came to the town each spoke a different language and had different customs. As well as learning their languages, he took great interest in identifying the tribes from the looks and dress of the people. He respected the different greetings and manners of handshake, which were most important for the natives. Some handshakes were quite elaborate, and it became a sort of a game, rather like those he had played as a child. The handshake was only the beginning of the greeting, and although the languages were different the overall process was the same. The greeting words were 'habari yako' ('how is your news?') or the equivalent in each of the languages. Commenting on the weather, or asking what town a person came from, was just not the done thing. The object of the greeting was to find out what tribe, or what family in the tribe, a person belonged to (that is, were they friend or foe?). So the greeting would continue: 'How is the news of your family?' 'Fine' ('mazuri') or just an 'ahaa' was the reply.

Learning from freed slaves

'How is the news of your cousins?', 'Ahaa' and so on, sometimes for minutes. Once one was sure that the other person was 'OK' the real news would start: 'However, my brother died last week!' Greetings took time and delivered a lot of information if you knew how to read it.

This was the start of Martin's continuous quest for information about the native tribes. Although there were many tribes and sub-tribes, he eventually identified the five main ones. These were the Kikuyu, Maasai, Luhya, Luo and Swahili, and each had several sub- or associated tribes.

The Kikuyu were of darker complexion and their language was Bantu-based. They were agriculturists and also animal herders. Martin found out that, traditionally, the women tended the crops and the men the cattle, sheep and goats. The extended family plot of land was an important possession and the basis of their livelihood. In his future travels through Kikuyuland he would be impressed by the work the women did and the loads they carried. Some of the older women had an indented band across their foreheads where a hide strap had used to be tied. The strap would be attached to a bag that supported heavy loads such as fire wood and water. The women were, like many Kikuyu, hard-working and industrious. He remembered once thinking that a Kikuyu woman, bent forward carrying a heavy load of sticks up a hill, resembled an amazing ant bearing a stone several times its own size. The Kikuyu had religious beliefs and a culture of government by councils of elders. However, each section of the clan was relatively independent, so an agreement made with one chief did not necessarily hold with another.

Opening Africa

The Maasai had a generally redder complexion and straighter facial features. They were mainly semi-nomadic and herded their cattle from one pasture area to another. Some of them, like the Chagga, were more agriculturist than nomad. Their cattle were sacred to them and were seldom killed as they were a symbol of wealth. Cattle also provided the Maasai with vital vitamins. Martin had sometimes been offered blood from a cow's neck to drink. A slit was made in a vein in the neck and the blood collected in a vessel. The slit was sealed by being pressed tightly once enough blood had been taken. This way the Maasai could continually drink fresh blood.

In many ways their whole lives revolved around the needs of the cattle. In the rainy season, when the plains were lush and green, the cattle would be put out to graze. Towards the end of the rains the tribe would move with their cattle towards the rivers that provided water for the dryer time of the year. This cycle of annual migration was shared with the millions of wildlife, like the wildebeest, zebras and the lions that followed them. The Maasai needed a vast area of land so that their cattle could move around. If a path was blocked by a farm it would mean trouble. Hence the Maasai concept of land ownership was different from that of the other agriculturist tribes. This meant that there was a potential for conflict with land-fixed tribes like the Kikuyu or Luo.

A Maasai group ('kraal') was a polygamous family compound. About twenty of these kraals made up a village or 'boma'. The males were divided into three groups according to age: youths, warriors ('moran'), and elders. When youths became warriors, they moved to

Learning from freed slaves

a different type of village, called a 'manyatta'. In the manyatta lived the warriors, their mothers and sisters and uninitiated girls. In contrast, the kraal was made up of families of married elders. The older male warriors went through a long process (up to six months) of handing over the warrior status to the younger incomers, before they got married and joined the elders. There was great peer pressure on the younger ones to prove that they were suitable warriors and it was during this time that most fighting broke out with neighbouring tribes. The young men had relative autonomy within the tribe and in these times the elders had little influence on them. Understanding these dynamics – that they were mostly 'young lads' on the rampage – helped Martin in interacting with and influencing them. The Maasai warriors had a reputation for being fierce fighters, but Martin got to know them as straightforward, humorous people.

Luhya tribal life revolved around the extended family where polygamy was the norm, and this was fascinating to any European. There was a strict hierarchy among the men of the tribe; women had little status and could not own land, but were seen as eventual 'possessions' of their future husbands. When a Luhya inhabitant of the town died, all his fellow tribesmen came and celebrated around his hut for what seemed like days. They went into the nearby forest and uprooted a large tree, and buried him under the spot. They then went on to plant a new, smaller tree next to the uprooted one.

The Luo were another tribe Martin managed to get to know at this time and about which he would learn more about in the future. He found them a versatile and industrious folk who were quick to smile and often liked

Opening Africa

to play music and dance.

One day a young Luo arrived at the Freretown camp looking rather poorly. He had been snatched from his village near Lake Victoria, in the dead of night, by a group of native raiders. Although he was a brave fighter, when the 'night spirits', who wore strange masks, woke him up and pointed spears in his face, he had frozen in fear and been unable do anything. It was as if his whole will to act had drained out of him. He could not understand why he had behaved so passively – it was as if he was programmed to do so. He was tied up and marched for two weeks to an area further inland, where he was sold to Arab traders in exchange for a few cattle. The traders put him in chains and he walked the two-and-a-half-month journey to the coast carrying a load of ivory tusks.

On arrival, he and the other slaves that were still alive were put in a dark cave where they waited for a week. His friend who had been chained next to him had died halfway through the march. The slave trader just unhooked his chains in the morning and dragged the body away. The young Luo had never felt so sad, but there was nothing he could do. He was herded onto a dhow early one morning and chained to the galley below the deck. Two days into the sailing, he heard shouting as another ship came alongside. Sailors in blue coats boarded the dhow and came down to the galley, where they cut the slaves' chains and set them free. Within two days he had arrived in Freretown.

Martin asked the young man what his name was.

'Barak,' he replied, in a half-frightened voice.

Martin tried as usual to link the new word with

a similar-sounding Maltese one. He often found that he hit on a matching meaning, more as a joke then anything else.

'Ahlla Ei Berak,' he said, which was a common Maltese greeting-word that his mother always used, meaning 'God bless you'. Barak's eyes lit up and a smile came to his face – his first smile in a long time.

Martin got along well with the young man for the rest of his time in Freretown. He learnt a little of the Luo language, and something of the people's culture and their ways of fishing and working. The other man learnt the skills Martin taught him and slowly regained his self-respect.

Martin also got to know other tribes, and found that there were many similarities between them. However, they differed in details that were important to the individuals involved. The people lived in family groups, several of which made up the tribe. Although they tended not to have overall chiefs, the groups shared cultural similarities and taboos. The family was seen in a wide context to include not just husband and wife (or husband and wives in some of the tribes like the Luo and Luhya), children and parents, but also cousins and 'blood brothers'. The family bond was important.

Most tribes had a council of elders who chose a chief often after several rounds of voting. Once picked, the chief carried much authority and was always the best 'contact' in a tribe. There was a check on the chief's authority in the consultations among the elders over the decreeing of laws. The chief could in most tribes be deposed by the elders' council if he broke certain rules regarding drunkenness, quarrelling, overeating and governing without regard to the input of the elders. So there was a form of democracy.

These tribes had a long history. The more Martin learnt about it, the more interested he became and the more respect he had for the indigenous people. The tribes had migrated from different areas to 'discover' suitable regions in which to live, sometimes fighting to take them over. They had hierarchies among themselves, with one tribe looking down on another – usually the stronger looking down on the weaker. These things suggested why fighting began between tribes, but they didn't really explain why conflict persisted continually.

Taboos and superstitions were part of the culture. The 'medicine man' or 'witch doctor' was the figure that instilled the fear of 'night runners' or 'spirit men' who came and took people away. There was indeed always a real fear that, from one day to the next, they might be pulled into slavery.

Slavery was the very worst thing. Martin's first perceptions of Zanzibar, with its illegal but still active slave trade, were imprinted in his memory. He learnt about the evil politics of the trade carried on with the upcountry tribes, and understood that it poisoned their lives and their culture and was the major cause for mistrust between tribes.

The East African tribes had, for over 300 years, suffered from the slave trade. Trafficking had grown from 10,000 slaves a year in 1811 to about 40,000 in 1839, with an estimated peak of 70,000 slaves being sold in 1860. Although trading slaves was made illegal in 1873 by the Anglo-Zanzibar treaty, illegal trading continued into the 1880s. Slavery itself remained legal until 1897 in Pemba, until 1907 in the coast region and as late as 1921 – the start of the British mandate – in Tanganyika. The main traders in James Martin's time were the coast Arabs and the Swahili coast people (years before them it had been the Portuguese, Dutch and British, who operated mainly from the West coast of Africa). The Arabs and Swahili had over the years made Zanzibar their "safe base" from which they carried out the slave trading. Initially on the coast but they eventually had to venture into the very heart of Africa to plunder their slaves. The users of slaves, or the 'market', were the plantations (cloves, cotton, fruit, palm oil, grain and gum copra) in Zanzibar and elsewhere around the world, which depended on this labour to run their new industries. The Mombasa area employed almost 50,000 slaves in plantations at that time. Domestic slavery in Arabia and elsewhere was another market, as was sexual slavery, and slaves were used to transport ivory and ebony, wild animal skins and palm oil from inside Africa to the coast. The slave trade affected almost every tribe and depopulated large areas.

Quite often African tribes such as the Nyamwezi, who lived between Lake Victoria and Lake Rukwa, would capture people from their fellow tribes and make them slaves, trading with the Arab and Swahili traders or even with other tribal chiefs. Livingston had highlighted this problem in around 1866:

> Livingston asked the locals repeatedly why they found it necessary to sell their people to a handful of intruders, and was told: 'If so and so gives up selling so will we.' 'He is the greatest offender in the country, it is the fault of the Arabs who tempt us with fine clothes, powder, and guns.' 'I will fain keep all my people to cultivate more land, but my neighbour allows his people to kidnap mine

and I must have ammunition to defend them.'[2]

There was also the Kore tribe, which was defeated by the Maasai in a fight and all the survivors plundered by a Somali tribe and taken as slaves. They were eventually freed by the British who settled them on the island of Lamu.

The fear of slavery affected the whole psyche of the Africans and bred helplessness, apathy, hatred and inter-tribal wars, with the stronger tribes trying to kill and plunder the weaker. Kings or chiefs traded their subjects, conquerors traded their captives, courts traded those they had sentenced. This led to an inferiority complex that in turn made slavery easier. Prices recorded were $1 for a child, $12 for a beautiful young girl; at this time one horse sold for twelve slaves.[3]

Ivory was the other main trading commodity. The trade in itself was passable as long as there was a surplus of animals and the local people benefited from the meat. (Today's conservationists would disagree, but in those times too many elephants could be a menace.) However, slaves were often used to transport the ivory, and hence it was part of the vicious circle that developed in many tribes: cattle being stolen to pay for ivory that would be carried by slaves, either traded or captured (cattle were the internal currency to pay the natives, ivory and slaves the export goods taken by the foreigners).[4]

Although the trading of slaves had recently been made illegal, it was still legal to hold slaves. This made it very difficult for traders to be captured, unless they were caught red-handed. They only had to plead ownership for things to be legal (rather like being able to own cannabis in today's world but not to trade it). Trading was thus

Learning from freed slaves

driven underground, which made conditions for the slaves even worse. Transport in the interior was slow and by foot, so that once a caravan of slaves was taken they disappeared into the unknown and were difficult to find. The task of redeeming the scourge of slavery seemed almost impossible at the time.

These are all things Martin learnt at Freretown, and was to understand more of in the years to come. His time in Freretown gave him a unique introduction to the tribes, areas and languages of Africa. He was probably better versed in the different native languages then any European before him, even though he was illiterate. Learning made him less apprehensive of meeting the indigenous people and of what lay ahead in 'darkest Africa'. He admired the simplicity and humour of the people, with which he could connect. This was the start of a quest for knowledge that would endure until he left Africa, as would the feeling that the assistance he gave them would need to continue.

5

Discovering the Maasailand with Joseph Thomson, FRGS (1883–1884)

Every month or two Martin made the journey by dhow to visit the CMS Mission built on the site previously occupied by the slave market; the high altar of the cathedral stood over the spot where the whipping post had once been. He also liked to renew acquaintance with Lieutenant Lloyd Mathews and with the crew of HMS *London*, whose captain had been murdered by slavers off the island of Pemba. He had been succeeded by Captain Brownrigg, and Martin now wished to meet the new senior naval officer in order to keep open a possible line of travel back to Europe. It was on one of his visits to Zanzibar that Martin met Joseph Thomson, the geologist and explorer. Thomson had arrived there from England on 26 January 1883, with the declared aim of opening up the direct route from Mombasa to Uganda.

    Africa was, of course, not completely unknown in Thomson's time, but the geography of central Africa had always been a mystery to Europeans, as also to the Greeks and Romans, and there was no knowledge of the Great Lakes or the source of the Nile river. David Livingstone, the first great explorer, crossed Africa along the line of the Zambesi (latitude approximately 10° S) in 1853 and 1856. Two years later, in 1858, Richard Burton and John Speke reached Lake Tanganyika (approximately 5° S).

Discovering the Maasailand with Thomson

Speke also in that year went north and sighted the Great Lake that he called Victoria Nyanza. He returned to the area in 1862 with James Grant and went up the western side of Lake Victoria to find its source at Rippon Falls (0.5° N of the equator). In 1864 Samuel Baker reached Luta Nzige, a lake north-west of Speke's Victoria Nyanza, and called it Lake Albert.

All the Great Lakes had thus been discovered to Europeans, but the extent to which they were linked up was not clear. It seemed conceivable that Lake Tanganyika could be the Nile's source. It was left to Henry Morton Stanley, in 1874–77, to link all these lakes and define Lake Victoria as the source of the Nile. However, Stanley circumnavigated Lake Victoria and passed by the western shores of the Kavirondo region without visiting it. Hence at this time the whole area to the east of Lake Victoria, from latitude 5° S up to just above the equator, was still unknown. This was Thomson's area of exploration.

German explorers had been very active to the south and east of Mount Kilimanjaro, signing treaties with local African chiefs by which their territories were placed under the protection of the German flag, and Thomson wanted to forestall them to the north. The unexplored area that was the target of Thomson's expedition lay between Mount Kilimanjaro and Lake Victoria and included territory peopled by the wild and bellicose Maasai, feared by all who ventured near them. Their country had in fact been scrupulously avoided by Arab slavers and traders. A few years earlier, an entire caravan led by an Arab named Mbaruk had been massacred by Maasai. Close to the Maasai lived the unpredictable and equally dangerous Kikuyu tribe.

Thomson had planned his expedition some years

before. He had been in Zanzibar in 1881, and had then returned to England to seek backing for his venture (which he got from the Royal Geographical Society) and to buy the necessary stores and equipment. He had decided to travel inland in East Africa without any white companions, but as he started to organize his caravan in Zanzibar he came across numerous problems. At this juncture he met James Martin and was impressed by his knowledge of the coastal people and his command of their languages. Martin had very good certificates of character reliability from the British Consul and the CMS Mission, and no doubt Lloyd Mathews and the officers of HMS *London* also helped. So Thomson engaged Martin as his second-in-command and put him to work to find the porters for their caravan. Martin seems to have had no binding ties with the CMS Mission at Freretown to prevent him from changing his job, and he threw himself into this new appointment with enthusiasm. It was to turn out a great success and, thanks to Thomson's lavish praise of him, to build him up into an incomparable safari caravan organizer and leader.

At the end of January 1883, Thomson went over to the mainland at Pangani with Martin to reconnoitre the initial stages of their forthcoming safari. They walked for six hours along the coast to Mombasa, and Thomson was disappointed to see that Martin was not a good walker and had skinned his heels badly. This did not augur well for the future, but on this occasion it was put down to badly fitting shoes. The numerous stores for the journey were brought to a depot at Tanga, which had a good harbour for dhows, and Martin came into his own steering the dhows along the coastal waters and in and out of the tricky coral reefs. It was not until 3 March 1883 that the caravan actually set off from Zanzibar, assembled on the mainland opposite Rabia (nearer Mombasa) and, with all the fuss and cacophony of such a large and excited gathering of 'askaris' and hangers-on, commenced its exploration safari marching behind the Union Jack and the red flag of the Sultan of Zanzibar.

There were over a hundred men in Thomson's 'Victoria Nyanza and Mount Kenya Expedition', and three donkeys to carry the sick or injured if the occasion arose. The porters carried loads weighing the regulation sixty-five to seventy pounds; these were balanced on their heads, and it was uncanny to observe how the men moved, swerved, climbed and even ran without unbalancing their loads. Nearly half of the baggage contained items to be used as gifts for the natives – thirty-four loads of iron, brass or copper wire, fourteen loads of bales of cloth, and as many as twenty-nine containers of beads. Then there were the tents, tent furniture, clothing, boots and personal stores, including scientific equipment and five cases of ammunition. (Although some dry provisions were also carried, the main supply of food was expected to come from wild animals shot en route.) As the contents were used up, the loads would be lightened uniformly. The porters received three months' pay on departure, and there was the strong temptation, in the initial stages of the safari, for them to defect from the lines, with or without their load, especially at night-time. To prevent this, the caravan included a dozen armed soldiers – the askaris – who also provided the column's security against wild animals and aggressive natives.

Thomson was in no hurry. He first wanted to visit the chiefs of the populated areas at the foot of Mount Kilimanjaro. These lay in the path of the caravans –

missionaries, explorers, traders and slavers who had previously travelled inland to Central Africa and round the back of Lake Victoria to Ruanda and Uganda and were well-known. Thomson thus decided it would be worth his while to cultivate their friendship and obtain as much information as he could about the Maasai, the Kikuyu and their territory. Furthermore, a British-presence in the disputed area south of the mountain, as well as being in that delicate period, could not but dampen the competitive tendencies of the German explorers.

The caravan moved slowly to Taveta and stayed there for some time. In May the first move was made to enter the uncharted country of the Maasai, but they were soon warned that their path ahead was blocked by 2,000 warriors on the warpath. Thomson retreated rapidly to Taveta. After consulting Martin, he decided to travel back to Mombasa to fetch more porters and askaris and additional stores, leaving Martin to look after the bulk of the caravan and the equipment during his absence. Left to his own devices, Martin lost no time in communicating with the surrounding chiefs and exerting his charm. He became a blood brother of some of them and won favours from others. He made friendships that would prove helpful on future safaris. He bought a sizable plot of land at Taveta by the side of a stream, and built a baraza, or assembly hall, in the centre with a flag pole flying the British flag. Near by, he constructed a dwelling for himself and a series of huts for his askaris and safari leaders and porters.

When Thomson returned to Taveta he declared himself astonished and full of admiration when he was led into the centre of a pretty, rustic village where before had stood the rankest jungle. He could hardly believe that these were his quarters and that the whole transformation was Martin's work: 'I was soon conducted inside our cosy grass built house, and while refreshing the outer and inner man, I listened with intense interest to Martin's tale of trial and trouble.'

Martin had in fact had some difficulty with Chief Mandara, the head of the Wa-Chagga clan, who had the reputation of being savage and bloodthirsty. While they were confined to camp at Taveta during Thomson's absence in Mombasa, food reserves had run alarmingly low. Mandara, learning this, invited Martin to visit him, but Martin was wary. He first sent one of his headmen, who happily returned with some food; then he summoned courage to go himself. He found Mandara to be most approachable and friendly, and the life of the little camp became much easier. It was through Mandara that Martin obtained a plot of land on which to construct the stockade that so impressed Thomson on his return from Mombasa. So it was natural for Thomson to invite Chief Mandara to visit the expedition's camp and to present him with gifts.

The month of fasting – Ramadhan – had now ended. It was already August, and it was time to face up to the journey into Maasailand. Hundreds of coloured beads had been strung into the customary length for the Maasai (any other lengths would have been rejected), and cloth had been cut and sewn in the manner they found acceptable. This time there were no alarms, and it seemed that the previous risk of a confrontation with Maasai Moran had been invented with the aim of bringing the expedition back to Taveta. While the Maasai women scurried out of their way, the older men were pleased to meet Thomson and Martin and raised no objection to their proceeding with the journey towards Victoria Nyanza, provided that they did not harm the cattle, caused no trouble and travelled in peace.

Gifts of beads, wire and cloth had to be distributed as 'chango' – a kind of transit fee – gauged according to the importance of the Maasai chief encountered. The Europeans in the caravan were the subject of much curiosity, and frequently had to suffer embarrassing personal inspection by the 'Moran' – the arrogant and contemptuous teenagers of the Maasai, who roamed the countryside in aggressive spear-waving groups, fearless of any opponent, until they had blooded their spears. The clothes and boots of the intruders were handled with guffaws and frivolity, hats knocked awry, and efforts made to snatch any loose object or any unattended load.

So the expedition proceeded: rising early to be on the march, with Thomson leading and Martin bringing up the rear; a halt after two hours for a brief rest; a meeting with Maasai tending their large herds of cattle or a chat with little groups who gathered around the strangers (the inevitable Maasai greeting of spitting on introduction, on agreeing to something and on departure is their expression of goodwill). Then on with the journey: at times tough and slow, at times covering twenty miles a day, and on one special occasion no less than seventy miles in one day. Obstacles that arose in their path could be formidable: rivers to be bridged; mountains to be climbed or circumvented; dense forests to be cut through; ravines to be traversed; steep escarpments to be negotiated; interminable thorn bushes to be avoided; days of waterless and shadeless desert to be crossed; and, not least, the constant threat from unpredictable wild game. Martin washed his face when adequate water was available and cleaned his teeth with a six-inch stick of freshly cut bush twig in the local manner, chewing the edges to expose the fibre which acted as a brush.

He never ceased to be amazed at the beauty of the vast open spaces. With the daily hard trek and enthusiastic organization that occupied his whole being, he particularly savoured the relaxation of dusk and dawn. The rising of the sun, which happens so quickly in the tropics, was the best part of the day. The sun comes in horizontally and lights up the vertical lines of trees, animals and people. They are clearer and crisper, and the light brings out all the subtle tones that are lost when the hot sun is overhead and melts the trees into the surrounding ground, blanking or blending out the natural colours as on overexposed film; the vertical lines are lost in the onslaught of heat and light of the African sun. Faces appear mellower while the dawn lasts.

There was an abundance of game in the plains of Maasailand. Besides zebra, wildebeest and impala, there were hundreds of smaller gazelle. One of the more beautiful and graceful of these caught Thomson's eye. It was small, about eighty centimetres tall, a third the size of a zebra, and had a light brown coat with a white underside and a distinctive black stripe. The animal's horns were long, ribbed and pointed, with an elegant gentle curve. The white patch on the rump extended to underneath the tail. A noticeable behaviour was the gazelle's bounding leap, which it used to startle predators. The creatures ate the low vegetation and grass of the savannah grasslands and their water intake came mostly from the grass they ate. They would be named after Thomson himself: *Gazella thomsoni*, or Thomson's gazelle. Animals were shot to provide food for the caravan, but it always seemed

Opening Africa

a shame when one of these creatures was killed. They could be semi-tame, and were often seen eating grass close to the huts that Martin would later establish.

Thomson had decided to go on a diversion to Mount Kenya and left Martin in charge of the safari for a couple of weeks. After inspecting Mount Kenya, Thomson discovered the falls that are named after him. He then rejoined the caravan, going on to discover Lake Baringo. His expedition then continued to Kavirondo and reached the town of Mumias (or Kwa Sundu). It then proceeded a short distance south west to reach the eastern shores of Lake Victoria. (This part of the journey would have passed just north of the present named town of Ndori, the home land of Barack Obama's ancestors.)

The scene when they reached the lake was described by Thompson in his book 'Through Masailand':

> An hour's feverish tramp, almost breaking into a run, served to bring us to the edge of Lake Victoria Nyanza, and soon we were joyously drinking deep draughts of its waters, while the men ran in knee-deep, firing their guns and splashing about like madmen.

The main objective of the expedition set for him by the Royal Geographcal Society had now been achieved, he had discovered the last remaining important secrets of central Africa, knowing this, Thomson, full of pride, gave a speech to his group 'on heroic lines more commonly heard at a city banquet or Mutual Admiration Society than in central Africa'. Their duty carried out, they then turned back and proceeded to the nearby village of the second chief of Samia (part of Kavirondo).

6

The 'awakening' of Obama's Luo tribe

That chief of a Luo village living in the region near the north-eastern shores of Lake Victoria was roused from sleep one morning by the sound of excited children shouting: 'A caravan is coming! A caravan is coming!' He quickly dressed and went outside to make preparations for this historic encounter.

He was a 'ruodhi' (chief) of several families of Luo's eastern region of the Great Lake, also known as Kavirondo. The Luo tribe had an old history dating back hundreds of years. The tribe had left their original lands in Wau in southern Sudan some 350 years previously, and over many years had moved southwards through the territory of Uganda. As they were primarily fishermen, they followed the different rivers and lakes looking for better fishing grounds. They moved at times when the fishing was bad or when their village was struck by a crop failure and the land became barren. They also moved to flee from enemies, and some groups moved when the land became overpopulated or fractions within the tribe quarrelled. The Luo were not nomads, but nor were they fixed to the land; the Nile was for many of them their guiding route. Some groups settled beside the rivers and lakes on the way south, some moved eastward and set up related tribes there, and some integrated with the indigenous Bantu tribes and lost their language.

Eventually they came upon Lake Victoria, the source of the great River Nile, several hundred years before the first white man had arrived there. It was a warrior chief named Ramogi Ajwang who eventually led them into the area where they now lived around the north and east and eventually also the south of the Great Lake. So their language of Luo (Nilotic) was spoken, in various dialects and by different branches of the tribe, from the south of Lake Victoria right up to southern Sudan, a stretch of over 1,000 kilometres. The original ancestors who found the lake must have believed it to be the promised land for fishermen, with its vast expanse and abundance of fish – a far cry from the original limited fishing rivers of Wau. The Luo had survived, and expanded and grew stronger in those times.

Once there, they did not move on. They were rightly proud of their heritage and way of life. Although they were fishermen, they had learnt how to keep fruit trees, cultivate crops and herd cattle and other animals.

However they were most proud of their skills in metal and blacksmith work. Samia was renowned for this expertise. Thomson:

> We found numerous smelting works, the ore being brought from regular mines, in a range of hills in the north. They smelted it in open furnaces of charcoal, heaped up against a low wall, at the bottom of which is a hole and a drain leading from it to carry slag. The

blast is kept up by a double bellows, worked with astonishing dexterity by a man standing. A whole day is employed in smelting the ore, and a mass of 15lbs to 20 lbs is the result. The moment it is thought to be ready, they turn out the hot mass and as speedily as possible cut pieces off with axes, dealing with great rapidity, Herculean strokes. The iron thus produced is first class, and the Wa-Kavirondo (Luo) especially those in Samia, are remarkable clever blacksmiths.

They made rings of beautifully polished metal worn by the young around their necks and legs. They also made spears, arrow heads, hammers, hoes and other weapons and utensils, which were used all around their country. A level of 'industrial' expertise not seen anywhere else in central africa.

A lot of their success was due to this diversity and ability to learn new occupations, and a willingness to take risks and explore new areas. They had a healthy understanding and respect for the environment. They were versatile and thus more creative then their neighbours, the Bantu-speaking Luhya and Kikuyu tribes, who were agriculturists, and also the hunter-gatherer tribes of the north.

All, however, was not well. Although there was relative prosperity as regards food, there were constant fights with other Luo groups, with their neighbours of Bantu origin and with the Maasai. The curse of slavery had been a constant threat, robbing them of good people, so the tribe's survival was a constant struggle. Disease, fighting and sometimes famine were their enemies.

Through the years the Luo had preserved their rich culture and had added to it. They had a tight-knit feudal-type society. Beside the chief there was the council of elders, who formed a sort of democratic government with laws, traditions, music, dancing and rituals. Next to the elders was the witch doctor, who took a special place in the village hierarchy. He wielded his own power over all the tribe – sometimes good, sometimes bad, but essential to the social dynamics. There were other layers of status in the tribe: the warriors, the fishermen and the farmers, and at the bottom came the women, who were all automatically lower then the men.

The women, however, had their own hierarchy: the chief's first wife coming first, followed by the others. It is often thought that the women were of no use, with no power, but in fact they did have their indirect ways of influencing what happened in the village. They formed their own group and every two months or so about fifteen of them would go in a noisy convoy to 'terrorize' the neighbouring village for a couple of days. At times like that the men feared this grouping of women and would not dare to approach them alone. It was as if they took all their frustrations out on the men in the next village, by clubbing up with their women and turning the usual social order upside down. Their weapons were their tongues and their bodies. Apart from this, women had a lot of work to do, cooking, collecting water and wood, tending the farm and so on. They would be married off, often with not much choice, to a neighbouring village or tribe. (Incest was totally taboo.) Women were part of their father's wealth, and were exchanged for cattle. They were lucky in that there were no circumcision rites in the Luo tribe; this put them apart from women in many other tribes, although they were not aware of it.

*****

The chief who had been woken by the children's cries that morning lived in a 'boma' of several huts, with his clan or extended family, cousins and aunts. The tribe was polygamous and he had four wives, with a hut for each wife and family. The huts were crowded; the whole family slept together on the flattened earth floor on thin woven raffia mats. In the day the huts were like a busy play area for the children, especially when it was too hot or when it was raining outside.

Today the chief instructed them all to stay in their huts and not go out to tend the crops. The boys were also told that there would be no fishing in the lake. Everyone was on the alert. He could not be too careful. Caravans brought gifts but also risks, and he had to minimize any danger to his people. He summoned up his drummers, and the young warriors of the village, who lived in separate huts, gathered with him to prepare to meet the caravan. They did not know what to expect. Some of the Arab trader caravans brought unrest to the area and the chief was apprehensive and worried that trouble was coming. He always greeted strangers with kindness – that was the tradition – but experience had taught them to be cautious too.

It was this tradition that the chief represented when he put on his official headdress and animal skins and took up his staff of authority to meet the visitors. As he waited with his warriors at his side and the drummers flanking the path, he cast his eyes to the sky and mentally took stock of his people. The headdress, although only made of feathers, seemed to weigh heavily on his shoulders. He knew where his people had come from but what would the future hold for them? The answer lay hidden in the minds of his own people, but he did not know how to unlock it.

He saw their minds as consisting of four zones. People tended to live in one of these at a time, switching from one to another.

The first was the ancestor zone, where the traditions and knowledge of the people were stored up. Each tribe member was taught these things that concerned all areas of their lives, from the rituals of birth to adult initiation, marriage and death, with their songs and dances and the social rules that adhered to them. They all had a duty to pass these things on to the next generation, as nothing was written down or otherwise recorded. This ancestral knowledge formed the identity of a person. If ever a man was asked who he was, the inevitable reply would be 'the son of Akongo', or whoever. The name identified not only the person but also the trials and fortunes of the tribe, who their friends were and who their enemies had been; it became a passport for acceptance and a welcome sign. It closed the door to enemies and made them take a step back. If people's ancestors told them of danger, or not to associate with someone, they would instinctively take heed. So the ancestor mind zone created divisions between tribes. It told them where they had come from and provided a code by which to survive the future.

The next mind zone was the social zone of living a normal, happy life together. Much of people's time was spent talking to one another; the boma offered constant contact. There was always some intrigue or other, always fun. Laughter was the sign that people were living in this zone, and sometimes the poorest people laughed the loudest. They would here, enjoy their social dances and music which gave them all, young and old, so much fun.

This zone was where the knowledge of how to cook and prepare food came in. The women would spend most of the day preparing the evening meal that they would all have together. The fish caught early in the morning in the woven basket-like nets were given over to the women – silver perch, rounded more than flat, between one and two feet long. Gutted and cleaned, these would later be slit under the belly and opened up like a book. Sticks would be put in the ground around the wood fire, at a slight distance, to regulate the heat. The fish would be staked between the sticks with the skin on the outside and the soft flesh closest to the fire to cook. Once cooked, the fish would be taken off the stakes, filled with herbs and some vegetables, closed up and thrown on top of the wood cinders on the edge of the fire. Once the skin was burnt and had formed a crust, the meal was ready. The ashes were cleaned off and the fish were served on wooden plates.

Bananas grew everywhere and needed little maintenance; they were the staple diet. The women would go out into the fields and cut a branch of green bananas. Any ripe ones were thrown away, but usually they never got to be ripe The skin was peeled off using a sharp-edged stone as a knife, and the hard bananas were thrown into a pot of water collected from the river or the lake and boiled for at least an hour. Herbs were added. Once cooked, the bananas were put on a wooden plate and mashed up, with the addition of herbs again. The mashed bananas would be served warm.

The more skilful farmers grew corn or maize, which was crushed and ground with sticks in long, rounded-bottomed vessels. This was the basic ingredient of their 'bread'.

The social zone also carried the games of courtship and the intrigues of the young adults. A wife was bought with cattle, but usually it was the richer, older men who had the cattle and hence the wives. Young men, however, always found ways to survive, either by stealing cattle or by daring an intrigue with the next village. To have too many wives was asking for trouble.

The third area of the mind was the work zone. Here was all the knowledge of the skills of fishing, boating, farming and planting and above all, metal working. The knowledge of how to hunt, fight and use spears was also found here. This zone helped the tribe to live and survive. They had variety and choice in what they did, but when working they positioned their thoughts in the work zone of their minds.

The fourth and last, but by no means least, of these mind zones was that of the spirits.[5] This the chief believed to be his people's major negative social inheritance. The spirit zone was a powerful controller of the body. It was the area that the witch doctor had access to and he spent much time in developing and exploiting it. This could sometimes work for the good; for example, when the chief himself couldn't sleep he would ask his witch doctor for something against this sleeplessness (though he seldom liked to do so). The witch doctor would prescribe a 'dawa ya ku lala' ('medicine to sleep' in Swahili), which consisted of a few special bones tied together. It was given with a little ceremony and the pronouncement of a few words. Having tied the medicine by his bed, the chief had no further troubles and slept like a baby. The spirit zone was also used for tribe discipline and policing. If the witch doctor put his known sign (usually some chicken feathers and bones sprinkled with chicken blood) over the door of a hut, no one in the tribe would dare enter that hut for fear of acquiring a curse. (There were ways to circumvent this danger if one really needed to get in.) So this power was an effective mechanism for controlling the tribe that should not be overused. The witch doctor's work was effective and had its place.

On the other hand, the witch doctor occasionally put a spell on people, and for whatever reason (sometimes there were ulterior motives) he would solemnly proclaim to someone that they would die in two weeks' time. Following this a seemingly healthy man would waste away and in two weeks to the day would fall dead. The chief never understood this and did not like it, and he continually tried to curb these undeniable powers.

The witch doctor raved on about his ability to call down the masked night raiders that would come and take people away; this was the worst curse of all. So paradoxically the slave trade and the stealing of men by night enhanced the power of the witch doctor. It was sometimes thought that the capture of men had even been instigated against his own people by the witch doctor himself, but that was never proven. But it was known that other chiefs had themselves sold people of their own tribe to the slave traders.

When someone was in this mind zone they could become paralysed by fear and unable to control

Opening Africa

themselves. When one of the tribe sat in an apathetic stupor not doing anything, it was a sign that he or she was in this state of mind. The effect was not always bad, however: it could turn men into fearless warriors, and young men and women going through painful initiation rites or illnesses could be freed of pain. But it could also invoke senseless, out-of-character, impulsive vengeance and slaughter. The witch doctor would use every opportunity to increase his hold of the subconscious part of the mind through his rites and ceremonies, especially with the mothers and their children.

Overall the chief knew this was a negative inheritance that hindered his people. Things had to change. He prayed to be able to reduce this vulnerability as well to combat the other threats of illness, violence and abuse that they faced.

*****

The beating of drums brought the chief back to the moment and he saw the caravan approaching, they were making their way back from the lake area. It was led by two white men, the like of whom his village had never seen before. One of them, who had a large and impressive black beard, walked right up to him and greeted him in the Luo language. He shook the chief's hand hard; at the same time he put his left hand under the chief's forearm, in the way of showing respect to one's elders. There seemed to be an instant acceptance of this short, bearded man. How did he know the Luo greeting, when even the chief's Nandi neighbours never greeted him that way?

Martin then introduced the 'second white man' as his 'chief', Thomson, to the Luo. He had a lighter white

The 'awakening' of Obama's Luo tribe

complection, moustache and was taller with a peaked white hat. The Luo chief had heard about the effort of the Europeans to stop slave trading and he was in full support. Hearing it directly from Thomson and Martin gave him hope for the future. The white men brought gifts of beads and wire for each of the chief's wives and items for himself. A brief handshake for the two men was a giant leap for the two civilizations.

The chief invited the expedition to camp in the area. 'Next day', Thomson writes:

> Finding ourselves among very pleasant people, we laid aside our natural reserve, and, pocketing our high dignity, we set all the young people of the village to trip it. In the cool of the evening Matin and I illustrated the 'poetry of motion' as practiced in Malta and Scotland: that is to say, Martin tried to initiate the damsels into the mysterious charm of the waltz, (probably the 'Maltija' dance) while I showed them how to do the 'fantastic' in the spirited movement of the Scotch dance. Need I say that Martin was simply nowhere, while they became enthusiastic over my performance. That night, as I sat musing and star-gazing, I concluded that the Wa-Kavirondo (Luo) were decidedly susceptible of civilizing influences!

There followed an incident in their camp where some of their goods were stolen from a tent by the locals. After a display of anger and threats from Thomson the goods were eventually all returned. This temporarily souwered the good feeling that had been created. However, as Thomson noted in his book:

> This adventure curiously enough seemed distinctly to raise us in the good-will of the people, and in the afternoon we were such excellent friends that they stood without fear to be photographed – Njemps being the only other place where this took place. The young women here were very well shaped, meriting as regards the figure the distinction of being called tall and handsome, though they are unusually narrow proportionately across the loins.

Thomson and Martin left a lasting impression, not only by being the first white men to go to the area but because it had seemed to the tribe that they weren't actually complete strangers. The Europeans were, in turn, impressed with the Luo people. The chief must have felt that this was a turning point for his tribe and that maybe his dream would come true: to move his tribe forward by reducing the fear and increasing prosperity and development. His optimism would be partly rewarded in the years ahead.

*****

Thomson and Martin's visit in 1884 opened up the boat route to Uganda (it was easier then walking the 150 miles or so to Kampala), and Martin would over the next

The 'awakening' of Obama's Luo tribe

ten years regularly pass through the region on his way to Entebbe. When larger steel boats were carried in overland and launched from Port Florence near Kisumu, more people visited the area. The caravans, some of which Martin led, brought trade and new ideas. The British further increased their presence in the region with the establishment of a station in Mumias by C.W. Hobley in 1895. Hobley himself visited the area in 1894 with Martin and again in 1896. Martin became a household name in the region and as a Government Administrator of the wider area was a kin to a governor of that region

Other areas were feeling the influence of the British administration. The British were attempting to stop the inter-tribe fighting and in 1896 a 'punitive expedition' was mounted in support of the Wanga ruler Mumia in Ugenya against the Umira Kager clan led by Gero, who were disturbing the peace. Over 200 men were quickly killed by a Maxim gun. In 1899 C.W. Hobley led an expedition against Sakwa, Seme and Uyoma locations in which, reportedly, 2,500 cattle and about 10,000 sheep and goats were captured. This fight against inter-tribal war and its causes – one of them being slavery – was seen by some as British oppression. Others welcomed the establishment of a common order that would create an environment in which the people could live in peace and develop.

In 1901 the railway reached Kisumu's Port Florence. Trains started operating in 1903, making the area more prosperous as all the Uganda trade went through this point. It became the railway terminus. Missionaries began to move into the area and set up schools. Here they targeted the 'spirit zone' of the local's minds to fill it with their teachings and slowly drive out fear. The

Opening Africa

Luo relationship with the British grew in a positive way, and to prove it, in 1900, the Luo chief Odera provided assistance to the British in the form of 1,500 porters to help carry goods.

Perhaps one of the most significant developments came in 1915 when the British government sent Odera Akang'o, the Luo ruoth (chief) of Gem, to Kampala, Uganda. He was shown the British settlement and the schools that had been set up, and was impressed by what he saw. Upon his return home he initiated the adoption of western styles of schooling, dress and general hygiene. This was the school system that started the development of the Luo people into the present era. Eventually the spirit zone of the tribe's mind was filled with western education and the witch doctor's influence was slowly pushed out. It was one of these first schools that Barack Obama's father attended in the 1920s.

*****

The journey back to the Mombasa coast from that first visit to Lake Victoria was meant to be quick for Thomson and Martin. The men were tired, having been on safari for over a year; their loads of gifts and stores were practically exhausted, and the objective of the expedition had been satisfactorily reached. Then disaster struck. Thomson was gored by a buffalo, and afterwards developed dysentery. The latter affected him badly and he was at death's door. For six weeks he struggled, at times losing consciousness, while Martin watched over him, nursing and comforting him. They had no medicines and there was little that could be done for the sick man out in the African bush. Martin took charge of the expedition, and had a difficult task maintaining discipline among the men, who refused to drag their feet indefinitely when they had thought their return home was so close. Gradually, Thomson regained his strength and, coaxed by his faithful companion, started the slow trek to the coast. The safari had lasted eighteen months. He had been a lucky man.

Thomson owed his life and much of the success of the expedition to Martin, and he could thereafter never speak too highly in his praise. In his book about his journey through Maasailand he speaks thus about Martin:

> I am happy to say that I had never reason to regret my decision to take him. Though unable to read or write he was very intelligent and could talk about ten languages in sailor fashion. In every respect, manners, language, dressing &c., he was far above the average sailor, and from the first I never scrupled to treat him as a companion. To show how well we got on, I might mention the possible unprecedented fact in African travelling, that we actually never once had an unpleasantness between us. If it were ever my lot to go back to Africa, I would seek for no better assistant.

Thomson and Martin had survived their first expedition into the heart of Africa. They had beaten the odds, as the statistics of European explorers at the time showed that the majority did not survive.

Malaria was perhaps the most dreaded disease and killed more Europeans in Africa than any other. In a Niger expedition in 1833–4, thirty-eight out of forty-seven (80 per cent) of the Britons in the party died of malaria. (The leader of the expedition, Richard Lander, was murdered.) As another example, all of Stanley's white companions died, probably of malaria, while searching for Livingstone. Although Livingstone proved in his journey of 1853–6 that quinine was a protection against the disease, he nevertheless suffered twenty-six attacks of malaria himself, and it probably contributed to his death in 1872 while he was still in Africa. If malaria did not kill, it left the sufferer unwell and debilitated: 'a mere skeleton, with brown yellow skin hanging in bags, his eyes protruding, his lips drawn away from his teeth' was the description Lady Burton gave of her husband after he had returned from Africa.[6] The majority of European explorers at this time did not leave Africa alive.

However, some people are born resistant to malaria and this resistance is inherited in their genes. It is known as the beta thalassaemia trait and occurs when an individual has one normal beta globin blood gene and one with a beta thalassaemia mutation. This trait status is thought not to cause health problems, although some women with the beta thalassaemia trait may have an increased tendency toward anaemia during pregnancy. Modern medical statistic tells us that it is most common in Mediterranean people of Italian descent and is an immunity derived from a culture's exposure to malaria at some previous time. It is also shared by other cultures, including some of those in Africa.

Hence, although it was not known at the time, the most suitable people to send to Africa would have been Italians or Maltese Italians. This is another example of how Martin's 'cultural DNA' made him appropriate for the job. Many a time he saw his colleagues come down with diseases, and he believed that he was charmed or always lucky, but it was probably all due to his Maltese genes.

*****

Surviving disease was not the only reason they succeeded in their trip, giving Thomson his 'finest hour'. A Scotsman from Nithsdale, just south of Glasgow, having studied geology at Edinburgh University, he was credited with having been the first to 'penetrate the Maasailand, inhabited by the most bloodthirsty people in the world' and then uncover the geographical unknowns that lay beyond: the Great Rift Valley, Mount Kenya, Lake Baringo, the north-eastern shores of Lake Victoria and the way to Uganda. These were the last pieces of the geographical central African jigsaw puzzle to be put into place. One of the remarkable things about this 3,000-mile expedition is that he did it 'without even firing a shot in anger'. Previous expeditions such as Stanley's and others had resulted in conflict and the killing of local people. Even Dr Gustav Fischer, a German naturalist, had attempted to do the same journey just prior to Thomson's trek, but he failed to penetrate Maasailand owing to Maasai hostility and sickness.

Why, then, was this duo, Thomson and Martin, able to succeed when others had failed for so long? It is fitting to call it a duo. Many writers have assumed that Martin had piloted Thomson through Maasailand and had been

Opening Africa

responsible for the success of the expedition. Thomson did not initially speak Swahili, which is where Martin helped out. Martin couldn't write where as Thomson excelled at it, resulting in a genuine partnership. As they never quarrelled and Thomson only had praise for Martin, it can be assumed that the success was due to a real team effort where both men played important parts and one would not have succeeded without the other.

The achievement as assessed in the late nineteenth century was the geographical or geological discovery. Today, 130 years later, it is clear that the expedition's success lay in the human element of bringing people and civilizations together and allowing a new area of development to begin.

So how did Thomson and Martin overcome all those hazards and find the 'last central African tribe to be discovered'?

The key is perhaps that both men shared a similar attitude and outlook. Both were 26 years old. Thomson was middle-class but came from a Scottish background which was still under the oppressive dark cloud of the Lowland and Highland clearances, when thousands of Scots were driven off their land and forced to emigrate. He would thus have developed a genuine respect for the underdog within the British (English) Empire and a non-confrontational approach to resolving problems. Hence he shared the same approach to the natives as Martin did and believed in the prevalent Scottish Protestant ethic that 'all men were equal'.

The 'awakening' of Obama's Luo tribe

The account of his meeting with the Maasai elders showed how Thomson operated. The Maasai had killed forty porters the previous year and he invited the elders emphasizing that no retaliation would be attempted. He described them as follows:

> They were magnificent specimens of their race, considerably over six feet, and with an aristocratic savage dignity that filled me with admiration.

The elders were assured that the past was forgotten and that they desired peace. If blood was shed, then Thomson's people would revenge themselves. He also added, 'Our caravan has a powerful white medicine man who would with the greatest of ease devastate the land with famine and disease, if you were to go back on your word.' (This was probably a contribution from Martin.)

A genuine equal debate was carried on with agreement on both sides. This cemented a trouble free passage.

On a number of occasions goods were stolen from the expedition, but Thomson never retaliated with force, but talked to the elders concerning the problem, usually with a positive outcome in returned goods and increased respect by the locals.

Thomson gives a strong sense of tact, prudence and good character in his book. He also took time to stop in different places for several days, to be able to get to know the people. Perhaps one of the masterpieces is the occasion when they met the Luo tribe and shared Maltese and Scottish folk dances with the locals. It must be the best way to introduce a culture and create mutual friendship from the start.

This then was no quick 'Indiana Jones style' three months in and three months out adventure, it was a prolonged journey of mutual discovery and all these aspects added up to significant breakthrough with the locals, which would result in a future pay back.

7

The partition of East Africa (1885–1886)

Once back in Zanzibar, Martin basked in the praise showered on him by Thomson. The Royal Geographical Society presented him with an inscribed gold watch for his work on the Thomson expedition, and his name surfaced whenever a new expedition within East Africa was mooted. First Lloyd Mathews, then Sir John Kirk wanted to hear detailed accounts of the long safari. They were interested not only in the opening up of the route to the Great Lake, but in the attitude of the African chiefs towards strangers en route, and towards representatives of large European powers bearing foreign flags. His Highness the Sultan also wished to hear Martin's account so that he could compare it with the various reports that had filtered through to him from his own subjects.

To all it appeared that the tribes at the foothills of Mount Kilimanjaro were being stirred by the active German missionary-explorers, notably Dr Carl Peters and Dr Juhlke. It was necessary to ensure that the influence of the Sultan (and therefore of Great Britain) did not suffer as a result. Lloyd Mathews decided to accelerate the build-up of the Sultan's military forces. He was given the rank of Brigadier General, mainly to impress the African chiefs on the mainland and so that he would have adequate rank and standing to deal with German officials. He engaged Martin as his second-in-command with the rank of Colonel (Martin never used the title, and to the men remained 'Bwana Martini'). The regular force soon numbered 1,400, and General Mathews could call upon a couple of thousand irregulars if they were needed.

The arms and drill training was done by Nubian or Sudanese non-commissioned officers, and Martin, who knew very little of military matters, exercised a general supervisory role, particularly in the quartermaster's department. Once a week, on Fridays, resplendent in the uniform specially concocted for him – a tunic laden with lots of gold braid, loose white trousers and a gold-handled curved sword, the whole surmounted by a large pith helmet bearing an imposing brass badge of the Sultan's Army – Martin reviewed the troops with General Mathews outside the Old Portuguese Fort near the Sultan's Palace. This was the Beit el-Ajaib, or House of Wonders – so named because of its unusual Victorian style of construction, with wide encircling wooden verandahs carried on slender iron pillars.

On 16 June 1885, General Mathews, with Martin and 200 soldiers, crossed over to the mainland on a twelve-day military expedition to the country around Mount Kilimanjaro. The purpose of the expedition was ostensibly to meet the request of Chief Mandara, made via Martin when he was there with Thomson's expedition, to be placed under the protection of the Sultan of Zanzibar, but it was also intended to fly the

The partition of East Africa (1855-1866)

British flag at a time when the African chiefs in the neighbourhood were being pressurized by Dr Carl Peters to opt for German overlordship. A German squadron of five warships was at the time cruising off the East African coast and had called at Zanzibar, and the orders given by Dr Peters to his colony-seeking officers were to act '*schnelles, kuehnes, ruecksichtloses*' ('fast, daring and ruthless'). Whoever first induced the local African chief to sign a declaration (which of course he could not read) was the victor. The chiefs were clearly perplexed, and at times signed agreements with both German and British representatives.

8

Caravan safaris (1885–1886)

Shortly after his return to Zanzibar, Martin was told that a newly arrived Englishman was looking for him. This turned out to be Sir Robert Harvey, who wanted Martin to organize a small caravan for a three-month safari along the Tana river, starting on the coast near Witu. There was no problem in getting leave from his military duties or in recruiting the required number of men for porterage and an armed escort. The task was accomplished to Harvey's full satisfaction, and Martin was back in Zanzibar in time to join a second military expedition under the command of General Mathews, bound for the Chagga country to drum up allegiance to the Sultan of Zanzibar.

All efforts to win over the African chiefs, even if so far successful, were soon to be thwarted by decisions taken at the highest political levels in London and Berlin. An international agreement between England and Germany would soon define the spheres of influence of the two European powers. The scramble for African colonies was at its height. Britain would eventually maintain the dominant interest in the vast area that was to be the future Kenya colony, while Germany would start to build her empire in Deutsch-Ostafrika by absorbing Mount Kilimanjaro, all the land at the foothills of the mountain, and all the land south of the British sphere, stretching to the border of Portuguese East Africa. The

Caravan safaris

deluded Sultan of Zanzibar, whose dominions allegedly encompassed the whole of East Africa, would be confined to the islands of Zanzibar and Pemba and a ten-mile strip along the Indian Ocean, carved out of the territory taken by the Germans and the British and over which he would have little effective control.

Back at base in Zanzibar, Martin found yet another request for him to organize a caravan to the interior. This time it was from Sir John Willoughby, a captain in the Royal Horse Guards and a personal friend of the Prince of Wales, the future King Edward VII. Captain Willoughby had heard in London of James Martin's exploits with Joseph Thomson, and was convinced that Martin was the best man to lead the private big game hunting expedition that he had decided to make in East Africa. His cabled enquiry (a cable station had been set up in Zanzibar only a few years before) was followed by his arrival by sea with his party.

Martin suggested to him that the expedition should centre around Chagga country and the borders with Maasailand. He knew the area and the people well and could confirm that game of all sorts abounded there. Because of the risks from the Maasai, Willoughby suggested for the supporting caravan an armed escort of sixty-five men, but on Martin's insistence this was increased by several more snider carbines and fifty muzzle-loaders. Captain Willoughby was clearly a wealthy customer and could afford the extra security. A large caravan on the march behind the British flag also fitted in with the official line (soon to be superseded) that stressed British supremacy in that disputed region. Martin himself was to have no fewer than five porters to carry his tent and personal baggage. Unusually, women

Opening Africa

bearers were included in the safari, though those with children were more of an encumbrance than a help, and there were also a number of slaves released temporarily by their masters on condition that they retained half their wages.

The expedition started off in November 1886 and in about a fortnight had reached the game area. The local chiefs were delighted to learn that Martin had returned and sent invitations for him to visit them. They each insisted that the expedition should pause in their own area. Doubtless they had an eye on the gifts and the trade that such a caravan brought with it. It was, of course, impossible to meet the wishes of all of them, and some, like Chief Mbugali, followed and caught up with Martin and complained, in a lengthy speech, that an old friend like Martin had passed by his place without stopping, even though a bullock had been killed in his honour. Martin proffered excuses and explained that the caravan was in a hurry to get to Taveta; he would call some other time.

At Taveta, Martin was able to guide the expedition to the compound and houses that he had constructed during his time with the Thomson expedition. Willoughby was impressed by the smart enclosure with its formidable thorn hedge and strong gateway. The square was enclosed on three sides, and on the fourth was a sparkling little river well stocked with fish. In the centre stood a tall pole, up which their flag was hoisted. Martin went to see his friend and blood brother, the powerful Chief Mandara, and was given a great welcome. Other chiefs and elders greeted Martin and showed their friendship by refusing to claim any 'chongo', or transit charge, from Willoughby.

Caravan safaris

The game hunt meanwhile proceeded. It was hardly to Martin's liking, as although he approved the shooting of game for food he did not like the indiscriminate killing of animals and was shocked by the massacre perpetrated by Willoughby's party. The final tally was over 300 beasts, 66 of them rhinos.

Martin had to guard against a different set of risks in this safari, not the dangers of unexplored territory or unknown tribes but the wild animals. Martin had by this time picked up a lot of knowledge about game. Hunting for the pot was easy, impala, thomson's gazelle and zebra were the preferred targets. It was just a question of getting close enough to the animal to get a good shot. The golden rule in stalking was to make sure that he walked into the wind, so the game ahead would not get his scent. Having been a sailor he was sensitive to picking up the wind direction, and instinctively kept the wind position in his mind at all times, even in just light breezes. He also got to know the tracks various animals left in the ground and how recent each track was. Information he learnt from his native trackers. It was the bigger game that was more dangerous, buffalo, lion, elephant, rhino and leopard. Even these were not difficult to hunt as the game only attacked if they were wounded or guarded their young. Usually they would take flight if they detected humans too close. To reduce the risks, every hunt had more then two or three hunters as a back up.

He still had the incident in his mind, when Thomson was gored by a buffalo. Buffalo were particularly dangerous as they had a habit of circling back on their path if they were being tracked and thus ambushing their pursuer. It was particularly bad if they were wounded. The wide horns across their heads also presented them with a protective shield when they charged. The biggest problem in this messy business was when the game was wounded and had to be tracked down and killed. It was an unwritten rule to always go after wounded game and relieve them of their suffering. Martin would have insisted that all the 'guest hunters' practiced their aim before the main hunt started, so as to avoid as few misses and wounded as possible. Once the game was shot the native trackers would prepare the trophies for taxidermy treatment and for mounting as a trophy. This in itself was a logistical problem, as the trophies had to be preserved in the right manner until they were treated. A taxidermist often accompanied the expedition.

Seeing the graceful movements and beauty of the animals extinguished, brought a feeling of regret to Martin, which more then cancelled out the initial basic excitement of the hunt.

The skills of managing his clients to keep them safe, happy and at the same time diplomatically 'manage' their lust for trophies, was a new challenge and is one of the many skills required of a 'white hunter'. The term 'white hunter' was coined eight years later (1894) in a hunting expedition led by Lord Delemare who referred to Alan Black as his 'White Hunter'. He is generally seen to be the first 'professional white hunter' in East Africa. Martin could, however, claim to have that distinction. This was to become a greatly sought after profession in East Africa in the years ahead and the dream career for many a young lad.

On the way back, Captain Willoughby took the opportunity of climbing the highest peak of Mount Kilimanjaro, and later met another big game safari on the move. This was led by Frederick Jackson and also had a large bag of trophies.

On the conclusion of the expedition, Willoughby expressed his deep gratitude and indebtedness to Martin for his excellent services. He gave him valuable commendations:

> I can thoroughly recommend him to anyone contemplating an expedition in this country. Indeed I consider any traveller who does obtain his services exceptionally fortunate, and in the best possible position for conducting a successful and pleasurable trip.

Willoughby's big game hunting safari had lasted six months, until May 1887. During this period the partition of East Africa between Great Britain and Germany was formally agreed, and much of the countryside where Willoughby had confidently waved the Union Jack became German territory. Of this he was probably unaware, and the new agreement could not have had any immediate practical effect either on the nomadic African tribes or on the wild game that roamed in search of its traditional grazing areas.

Following the return of Willoughby's safari, Martin rejoined the Sultan's army for another brief period before he was called upon to furnish and direct a porterage caravan for an eight-month safari by Sir Robert Harvey, with whom he had already travelled up the Tana river.

9

The long supply trek to Uganda (1889–1894)

> Upesi, upesi, hata Baringo
> - mbale kidogo, tutafika Uganda!
> (Swahili safari song:
> Hurry, hurry, we've reached Baringo
> - Its only a short distance to Uganda)

The Kingdom of Buganda stands at the north-west corner of Victoria Nyanza. Until the end of the nineteenth century it was accessible to strangers only if they approached up the west side of the Great Lake, which meant a long and unenticing journey from the coast. Approach from Sudan meant passage through the impassable Sudd, and negotiating hostile country and the warlike Maasai and Kikuyu tribes. There also existed a traditional belief that the first white man to enter Buganda from the east would overthrow the Kabakaship. Thus the Baganda for centuries had lived in splendid isolation. The curious thing is that they alone in East and Central Africa had developed an advanced system of government with a king and a royal court, ministers, a parliament, and a hierarchy of local and subordinate chiefs.

The king, or Kabaka, was absolutely ruthless and enforced his will without concern for human life. On his accession, every male relative in the kingdom who could in any way be considered to have the least possible claim to the throne was mercilessly murdered and thrown to the crocodiles in the lake outside the Kabaka's palace. In this way it was ensured that there were never any pretenders to the throne. Individuals, men or women, were shot or mutilated at his whim for petty indiscretions such as not wearing the correct dress or for speaking out of turn or giggling in court. When Speke, on his first visit to Buganda, presented a gift of rifles and guns to Kabaka Mutesa and explained their operation, the Kabaka selected one of the weapons, loaded it, and handing it full cock to an aide told him to go outside and shoot a man 'to see how it worked'. The order was obeyed without hesitation and without a thought for the victim.

The first recorded visit of a caravan to Buganda was that of some Arab traders in 1840. Next, in 1862, the explorers James Grant and John Hanning Speke, looking for the elusive source of the Nile, entered Buganda from the west and were detained for some months by the Kabaka before Speke discovered the source of the Nile at a cataract at the northern end of Lake Victoria. Here a large volume of water flowed out in unrestrained confusion on the start of its 4,000-mile journey to the Mediterranean. It was 28 July 1862. The mystery that had exercised the minds of the civilized world for so long had been solved. The famous Bernini fountain in Piazza Navona in Rome, representing the four great rivers of the world, shows a representation of the Nile as an African with his head hidden under a cloak.

Anglican missionaries in 1877 and White Fathers of Cardinal Lavigerie's missionary order in 1879 reached Buganda, again from the west, but the first attempt to enter the Buganda kingdom from the east met with disaster.

An entire caravan, with Bishop James Hannington and his CMS missionaries, was, on the express order of the Kabaka, massacred as it approached the Nile.

The pace of opening up what remained of unallocated Africa could not, however, be slowed. With East Africa now partitioned between Germany and Great Britain, there developed a race for mastery of Buganda. While the great powers in their European capitals haggled and discussed, in Africa they operated surreptitiously through 'imperial' trading companies such as the Imperial British East Africa Company (IBEA) and the German East Africa Company (GEA). These companies did indeed engage in trade, but they were probably more interested in furthering the expansionist ideas of their fatherlands, often going beyond the orders or the intentions of their respective governments.

In 1889, the IBEA Company appointed Frederick Jackson and Ernest Gedge to undertake a prolonged safari to survey, map and set up posts along the route to Uganda. At the same time, unbeknown to them, Dr Carl Peters set off on his own safari aiming to be the first to establish the Kaiser's influence over Kabaka Mutesa in Buganda. Dr Peters's caravan was not properly financed or organized. He marched up along the Tana river, brooking no interference and dealing savagely with any opposition. In Kikuyuland he shot six Africans who slowed his progress. This was Prussian ruthlessness at its worst.

Frederick Jackson's caravan, however, was organized with proper care and attention to detail. The IBEA was responsible for all arrangements, and the obvious choice for the leader of this large and well-equipped caravan could only be James Martin, who by

The long supply trek to Uganda

then had joined the IBEA Co. With Gedge he went to Zanzibar and chose his Zanzibari porters and askaris, many of whom had travelled with him on earlier safaris. Stores and equipment were purchased and a return made to Mombasa in time for the caravan to leave on 18 June 1889. Jackson remained behind, but caught up and joined the party a month later at Machakos. The caravan, over 500 strong and well armed, moved on for the next eight months in the country of the Maasai, the Kikuyu and the Nandi, avoiding warriors on the warpath and areas where tribal battles were being fought, but at times having to fight through tribal opposition. It is recorded that in this period Martin cut off his thick beard but kept his moustache, that Gedge and Jackson tried unsuccessfully to teach him to write, and that Martin kept tally of the men, loads and equipment under his charge by drawing a series of lines and hooks and his own concoction of hieroglyphs.

One thing the 1886 partition between German and British East Africa meant was that the usual caravan route to the interior, which used to go inland from Tanga and round Kilimanjaro (as Thomson had earlier), was now in German East Africa, so the caravan had to proceed inland from Mombasa, which was higher up the coast. A route through the Kenya region existed and was used by Arab and Swahili expeditions in search of slaves and ivory; however, this route went through the dreaded Taru desert.

The trek was along a narrow track of about 700 miles fraught with danger. All the luggage had to be carried by porters. Animals such as donkeys could not be used because of the tsetse fly, which infected and killed them. Each porter carried a load weighing up to sixty pounds for ten to fifteen miles a day. It was best to avoid the rainy seasons – either the short rains in October to November or the long rains in March to May.

Starting out from Mombasa and heading northwest the terrain was pleasant, with moderate temperatures tempered by the sea, although the high humidity could be oppressive. Vegetation included coconut trees, tamarind, oranges, bananas and mangoes. The first fifteen miles, or the first day's travel, were easy. The stop at Mariakindi marked the start of the Taru desert with its waterless terrain and red dust, sparse acacia and a few strange baobab trees. The fifty-five-mile desert took three days to cross.

The baobab tree is one of the world's strangest. Its fat, smooth-skinned trunk can grow in the most arid places. The trunk looks like a storage silo, and that is just what it is – storing up enough moisture, sometimes equivalent to gallons of water, to keep it alive. Its large fruit pods make vessels for the local women to carry water in. They also make good toy boats for children. The tree is considered sacred by the indigenous people and they would not dare to touch it. This suggests the possibility of a lost and thirsty traveller resting beneath the tree and pondering the option of hacking at the trunk to release the moisture that lies inside, and thereby killing the tree, or letting it live. Perhaps the ultimate environmental sacrifice was sometimes made, in which the traveller resisted harming the tree but died of thirst in its shade.

The expedition's goal was to reach Maungu, which was the next safe water stop. All their water had to be carried by the porters along with their full loads. Without shade the going was relentless. The heavy, precious water partly evaporated or was lost or even stolen along the way.

To cross this wilderness the caravans had worked out a method called 'terekeza', or 'two marches a day'. The first day, from Mariakani to Taru, was done in two walks with a break between. The food they ate was mainly locally grown fruit, such as mangoes and bananas. The second day's march started from Taru at dawn with the remaining water and food. They stopped at Buchuma for a rest and a drink. The caravan then pushed through until dark, when they stopped but without taking any food or water. Just before daylight on the final day, having eaten and drunk what remained, they pressed on with the objective of reaching Maungu hill. This they had to reach without water or food, so it was vital that they got there before nightfall, or they would probably die. The climb up the hill to reach the water that lay in a depression was a final challenge.[7]

A successful crossing of the desert required detailed logistical planning and total discipline from all. Most desertions of porters occurred in the early stages of the desert; roaming alone in this hostile environment usually led to death by dehydration. It was reported that Martin used to tie the porters together until they were far enough into the desert to prevent them from running away and perishing. The caravan was only as fast as its weakest link. Many men got sick and died. Disease claimed many (smallpox and malaria were common); attacks by animals (among them hyena and lion) claimed other lives.

> Sickness caused delay. It was not an uncommon occurrence, after 2 miles or so of a day's journey,

for a porter to break down. When this happened, the line of men was ordered to halt, a litter constructed, and five men's loads distributed among the other porters, an action likely to produce near rebellion (though the sick man's pay was distributed among them to prevent malingering). If the porter was desperately ill and water was far away, the caravan was split in two and the advanced party and invalid travelled at full speed to the source with the white man in attendance. After water was reached, the invalid and a few porters would stay beside it while the European and the other men returned to the rest of the caravan with water for those left behind.[8]

It was only when the caravan reached the Voi river (a hundred miles from the coast) and the Taita hills that the worst part of the journey was over and regular good water could be found along the way. The country was now open savannah with plenty of game. Impala and gazelle provided the main diet and vitamins at this stage.

Among other problems that affected the expedition was hostility from the local tribes through which the sometimes several-hundred-strong caravan passed. If discipline was lacking, some of the porters would steal crops or molest local women. The reaction could be severe, even with attacks on the caravans. Communication with local chiefs was the key and Martin was good at this. Negotiations agreed on a 'transit fee' that would compensate for these incidents. The caravan head porter, or 'kilangozi', would mete out to the culprits appropriate punishments that might appear drastic in today's world, but docking of pay or flogging or chain

The long supply trek to Uganda

gang duty was the law of the time.

The next major stopping point was Kibwezi, which was an eight-day trek. Along the way the expedition halted at Tsavo and Mtito Andei. Each stopping point presented its own trial for getting water. In places they had to dig into the often dried-up river bed to access the water below. Some pools were covered in weeds. It was always an act of faith that the water at the next stop would be all right. There was the added danger in this sector of lions attacking the porters at night. A ring of thorn bushes around the camp was necessary for additional protection.

On the twenty-sixth day of the trek they arrived at Machakos. The friendly chief was known to welcome travellers and it was one of the first posts that traded with the visitors. Jackson heard on arriving that the leader had unfortunately died the previous year, but his name, Machako, lived on. This was one of the larger posts at that time.

The next big stop was Ngong Bagas – now called Ngong – in Kikuyuland (this was near the area where Karen Blixen had her farm many years later). By this stage the party had been trekking for thirty days. The Kikuyu cultivated their land and were good trading partners so the caravan stocked up with fruit, vegetables and meat (from cattle and chickens). This had to last them until the next main food station at Mumias, which was three weeks away.

The journey climbed higher through the red-coloured, richly vegetated Kikuyuland, with its large forests. In contrast to the Taru desert, where temperatures reached 40°C, the highlands were often frosty with 0°C temperatures at night – a further challenge to the logistics of providing adequate clothing for each stage of the journey.

The next big event was the Rift Valley. They travelled along its ridge for a while, then descended 600 metres to the valley floor via the Kedong Valley. Once on the flat valley floor they passed the prominent volcano Longonot on their left and proceeded to the beautiful Lake Naivasha. Its covering of thousands of pink flamingos must have been uplifting for any traveller. This part of the journey was again easy going, with moderate temperatures and flat land. The caravan was by now 400 miles and thirty-four days into its journey, with another 220 miles to go to get to Mumias. The lake could be passed on the south-west side if the river on the east side happened to be flooded. The land provided meat in the form of impala and gazelles, which were abundant there.

They proceeded towards Lake Nakuru and then had a three-day climb up the other side of the Rift Valley. The journey through the Mau hills and their forests took another seven days. The Mau summit was a climb up to 3,000 metres, and again the temperature could get quite cold in the nights.

The next big obstacle was the thick tropical rain forest, thought to be one of the most hostile and impenetrable environments on earth, with tall straight trees giving a treetop canopy that cut out much of the sunlight. (It gave a feeling of awe, much like walking through a huge roofed man made structure.)There was, however, less undergrowth due to the diminished light, which facilitated walking. The air was humid and a potential disease carrier. There was a heightened danger of a different set of wild creatures especially snakes which including the giant python which grow up to twenty feet long. These are fearsome to behold but, in fact, move so slowly that they are less of a danger to humans than the smaller poisonous variety. The python caught, crushed and killed smaller animals and then swallowed them whole. The path way in the forest was often obscured which increased the travellers uncertainty.

Eventually they arrived at Mumias (or Kwi Sundu) at the base of Mount Elgon. The arrival of a caravan in Mumias was a special event, and it marked the end of the journey for most. Some of the porters would be dressed in their leopard skins and ostrich feathers, and a band of drummers would welcome them into the camp.

The Lake Victoria shores were rife with malaria, and Europeans were highly susceptible. At night clouds of mosquitoes hid the moon. The mosquito nets that were so vital at the coast were used again in this area. At this point they were fifty-one days from the coast. Probably several of the porters would have died by this time.

The remaining journey to Fort Kampala, or Mengo, would be accomplished either by boat along Lake Victoria or via another twenty-day march. In all, the Uganda caravan had been on foot for two and a half months, and once they had rested it would take them the same amount of time to return to the coast. The men on a caravan were away from their homes and family for between five and six months. Europeans saw their arrival back in Mombasa as a return to civilization.

This was James Martin's daily life, and it was no mean undertaking. He had to manage the porters of different tribes, see that they were all up in time in the mornings, provide for food, look after the sick, stop them diserting, limit the thieving, and keep track of the numerous goods and suplies they carried. No wonder that so few people had ventured into the interior and that Kenya and Uganda had remained cut off from the rest of the world for so long. When diseases like dysentery struck a caravan, porters could die by the dozen. The head porter had an important role, not only as regards discipline but also in terms of morale. One thing he would do when the trek was on a flat run would be to sing out a loud safari chant in a high (F major) voice:

'Funga Safarii! Quenda Mariakanii!'

The several hundred porters of different tribes would reply, with a loud and perfect (C major) pitch and timing:

'Eheee!'

It was awesome, and reflected the combined power of the caravan: a moment only Africa could bring. The head porter carried on and improvised on the content of the chant, often making comments about their difficulties or even about the white leader of the trek. Some comments were quite humorous and brought a smile to the porters' faces that the white leader himself never quite understood.

Jackson's caravan eventually reached Kavirondo and there found letters written months before, with an invitation from the Kabaka and the White Father missionaries in Rubaga for him to hasten to place Buganda under British protection. It was too late. Dr Peters had been there earlier, had opened the letter addressed to Jackson and had hurried post haste to the Kabaka and persuaded him, less than a month before, to sign a treaty of allegiance with Germany. Jackson was naturally upset, and disappointed that he had been beaten. He remained in Kampala for only a month – just enough time to rest and visit the CMS and the Catholic

The long supply trek to Uganda

missions, and to receive a return call from them – and then his caravan, led by Martin, started the return journey to Mombasa with news of the German achievement.

The safari back took three and a half months and reached Mombasa on 4 September 1890, but during the course of the journey the situation in respect of Uganda had changed completely. The highest representatives of the two European powers, regardless of what was happening in Africa, came to a mutually acceptable agreement whereby Great Britain handed over to Germany the small island of Heligoland situated in the North Sea off the German coast, and in return was given sovereignty over the African countries of Uganda and Equatoria and the small kingdom of Witu on the coast. In addition, Germany recognized Zanzibar as a British Protectorate. Dr Carl Peters had hardly been given time to announce his achievement before it was denounced by his government.

As Jackson's caravan made its way to Mombasa, another caravan, fitted out by the Imperial British East Africa Company and commanded by Captain F. Lugard, was moving westwards to build stations at Ngong and Dagoretti and await instructions. When these instructions arrived, they were to proceed forthwith to Buganda and there undertake the administration of the newly acquired country. Lugard arrived on Christmas Day 1890 and the next day made a treaty with Kabaka Mwanga.

The bringing of Buganda under the control of the Imperial British East Africa Company meant that supplies of all kinds would have to be sent regularly from Mombasa. Until the railway was built, there was only one way to solve this problem – the traditional foot safari, with ever increasing numbers of porters.

James Martin's worth and his position in the company as chief transport officer became more significant, and for the next few years he led important safari caravans to and from Uganda, a distance of 800 miles each way, on average more than twice a year. He had tremendous stamina, and seemed immune to the illnesses that beset travellers in the bush as a result of insects, wild beasts, poor diet and fatigue.

As he traversed the distance, he constantly sought to improve and shorten the route. With him lies the credit for discovering the more direct and shorter route to Uganda through Uasin Ghisu. The safari marches wound their way from camp site to camp site, or from post to post, these stopping-points often bearing the names of local leaders or chiefs. Some were well-built stockaded forts, like Dagoretti and Fort Smith; others were merely a clearing with a cluster of huts, like Mumias or Lubas, which helped to relieve the hardships of the caravans by supplying water and locally grown vegetables or trading tea and small essential goods.

Martin's first Uganda supply convoy left Mombasa on 8 December 1890 laden with stores, arms and ammunition for Captain Lugard. The 800-mile journey lasted three and a half months, and when it arrived the caravan was in a bad way, with several porters suffering from sores and ulcers. Lugard was shocked. He blamed Martin, and complained to the company that it had not allowed sufficient finance for the expedition to be properly prepared and conducted. Without much of a let-up Martin marched back to Mombasa, leaving Kampala in April, but he was back in Kampala on 9 December, bearing, besides a large supply of arms and ammunition, two horses and a pony and a steel boat in

The long supply trek to Uganda

sections that weighed seventy pounds. When assembled the vessel was twenty-four feet long and was named the *James Martin* (later, in 1907, it was part of the flotilla used to convey Winston Churchill on his journey down the Nile from the foot of the Murchison Falls to Lake Albert and on to Nimule in Sudan). With Martin on his 800-mile return safari to the coast in January 1892 went some 300 of Emin Pasha's Sudanese evacuees, who had been cut off by the Mahdi rebellion in Sudan and who were taking the long way back home via Mombasa, the Red Sea and Egypt.

The origins of the Sudanese troops in Uganda, whose descendants still form the backbone of the various branches of the African Rifle regiments in East Africa, go back to the time of the Mahdi's massacre of the British garrison under General Gordon in Khartoum in 1885. The Mahdi's victory cut off Equatoria from the north of Sudan and with it the Governor of Equatoria, Emin Pasha, and his garrison of around 10,000 Egyptians and Sudanese, which included a large following of women and children. Unable to travel towards Khartoum and with the way south along the river Nile blocked by the swampy and almost impenetrable Sudd, Emin Pasha and his army were 'lost'. An Emin Pasha Relief Committee was set up to finance and organize an expedition to find and rescue Emin.

The name of Joseph Thomson (who with James Martin had travelled through Maasailand in 1883) was first suggested as its leader, but the final choice fell on Henry Morton Stanley. Stanley, ever hungry for publicity and fame, mounted a formidable expedition from the west coast of Africa through Congo. In the event, Emin Pasha and his army did not need rescuing. The party gradually travelled down from Sudan to Wadelai and into Uganda. While Emin Pasha – who although employed by General Gordon had been born in the Prussian province of Silesia – went off to join the Germans in German East Africa, his Sudanese followers settled down in Uganda for good.

The Arab soldiers were the only trained military men in Central Africa, and from them the British newcomers to Uganda recruited their security forces and military units. These Sudanese troops were to give James Martin his first real headache at his new posting at Eldama Ravine, because it was near there that in September 1897 several hundred Sudanese troops started a revolt which spread as further garrisons mutinied. Three British officers had been murdered and numerous askaris had lost their lives before it was quelled.

Henry Stanley arrived in Zanzibar in February 1887 at the start of his expedition to find Emin Pasha while Martin was on a safari with Willoughby. Following the conclusion of his quest, Stanley left Zanzibar for the last time on 29 December 1889, while Martin was on his Uganda trek with Gredge and Jackson. Thus Martin never had the chance to meet one of the great explorers of the age, who had also spoken Swahili.

Over the next years, until 1894, Martin conducted large caravans to and from Uganda with amazing regularity (see Appendix 2). He became a well-known and popular figure on the safari that became more or less established along this regular route. He was allowed by the IBEA Co. to undertake trading on his own account, and according to reports was at this time making a profit of from £2,000 to £3,000 a year, mainly by trading ivory from elephant tusks.

He accompanied all the well-known leaders and

The long supply trek to Uganda

administrators who travelled to Uganda – Lord Lugard, Sir Frederick Jackson, Major J.R. MacDonald, Sir Gerald Portal and many others – and won their friendship on the three-month safari. He conveyed mail to Uganda and back, and with it his own verbal reports of historic political events that were unfolding in his path.

Some amusing anecdotes have been written about Martin at this time. Once at Tsavo river, hot and exhausted, the caravan porters stripped naked and plunged into the water. Martin was about to follow their example but saw some nude native women watching him with curiosity. He refused to undress in their presence, and although much in need of a bath preferred to wait until darkness fell before taking his turn. On another occasion, while he was traversing Maasai country with MacDonald, they were confronted by a group of spear-waving Maasai Moran, who insisted at all costs that MacDonald should give them some potent medicine to ensure protection and invincibility in battle. Not being able to satisfy their wishes, MacDonald was about to have them dispersed when Martin asked to be allowed to deal with the Maasai himself. Taking a bottle of Eno's Fruit Salts, he moved the Maasai a short distance away and, with some ceremony, poured water into tumblers to which he added a pinch of the effervescent salts. The effect on the Maasai of swallowing this cold 'boiling' medicine was impressive. They felt they were now impregnable, and readily rewarded Martin with four donkeys for his trouble. Martin explained to the amused MacDonald that he had not fooled them entirely since the Eno's salts could not but have a salutary effect on their bodies and thus make them fitter warriors!

10

Developing Ravine as a District Officer (1894–1904)

In June 1894, the British government took over the responsibilities, the obligations and the staff of the British East Africa Company, including James Martin, who suddenly found himself appointed a District Officer in the Uganda Administration. The new Uganda Protectorate extended its boundaries from the Mountains of the Moon in the west as far as Lake Naivasha in the east, in territory that is now part of Kenya. Its easternmost outpost was the stockaded camp of the District of Eldama at Ravine, forty-seven miles north of Nakuru. Martin was put in charge of Ravine Station and arrived there on 7 January 1895.

The area around Ravine and the safari trails leading to the station from Dagoretti in Kenya and from Mumias in the north were beset by raiding warriors of the fearsome Maasai and Nandi tribes, who when not fighting each other or other tribes attacked supply caravans, stealing loads and killing people. The trail being so vulnerable, the District Officer at Ravine had to provide armed escorts to the next post for the ever increasing number of private caravans run by missionaries and traders. The Nandi were particularly active in the area and had shortly before killed a British officer. Delay in mounting an armed expedition to punish the murderers merely encouraged them to be more audacious and arrogant, and reports came through to Martin that they were planning to attack Ravine. He at once set himself to reinforce the perimeter stockade, asked for reinforcements of askaris from other districts, and strove to build up his stores of food, arms and ammunition. He also asked for increased rations, and for trade goods such as beads and cloth to enable him to purchase more food locally: he believed in beads rather than bullets for solving disputes. The next station, Mumias, was depleted of men in order to help Ravine be properly defended.

Martin was no trained soldier, but he knew the country and habit of the people around him, and above all he was well informed. Because of his illiteracy, he privately employed a Goanese clerk, Anacleto de Silva, to do all his paperwork and to write his reports and letters, which he could now at least sign. He welcomed the change from the restless life he had been living, and discovered the joy and comfort of having his own bungalow furnished with all conveniences, many of which he had paid for himself. He extended warm hospitality to visitors, of whom there were quite a few as Ravine was an important staging post.

11

Founding a family and a city (1896)

Having now been in charge of Eldama Ravine for twelve months, James Martin found that he was a lot better off than he used to be. His part-time trading in ivory and beads had helped him build up his finances. He had been living in one fixed place now for a longer time than he had done for years, and he used this time to make his bungalow into a home. He wanted to build on this stability. He was now thirty-six years old and believed that it was time to start a family: it was now or never. Out on the treks he had always had something to do. Being in a fixed place he felt lonelier than he ever had before.

Since 1985 Eldama Ravine and Fort Smith had grown rapidly owing to the increase in traffic to the interior. It was estimated by Francis Hall that between 1,500 and 2,000 people would travel through on their way to Uganda between September and November 1894.[9] He was proud of the position he had established for himself and he was now well-known throughout the country.

Martin decided to go on leave back to the more civilized coast and get married. He left on 27 December 1895 for Mombasa and then, after selling off some of the goods he had brought with him, he went to meet his old friends in Zanzibar. While there he married, on 9 April 1896, a pretty petite Portuguese lady, Maria Augusta Elvira de Souza, the daughter of a respected doctor in

Founding a family and a city

Zanzibar. He had sometimes used the services of the doctor on his trips back to 'civilized Zanzibar' and had got to know him well. He also couldn't help noticing the doctor's daughter whom he had met several times. She had a slender figure and sharp features, and a darkish Mediterranean complexion not unlike those of some of the girls back in Malta. They were married by the British Agent and Consul, General A.H. Harding, and the witnesses were Arthur E. Raikes and Martin's old friend General Sir Lloyd Mathews. Martin's marriage certificate describes him as 'Transport Officer, Uganda, aged 36 years – son of Antonio Martini, a 3rd Officer in the Mercantile Marine'.

Martin took his time with a long honeymoon of several months' duration. While in Mombasa he met George Whitehouse, the chief engineer in charge of constructing the railway, who had arrived a few months earlier. Whitehouse had asked Martin to set up a Camp upcountry to recruit Africans for the construction of the railway. According to the studies carried out and Martin's advice, a central camp in the Kikuyuland would be the best option for both climatic and personnel reasons. Martin collected the required materials for the camp and took them up with him along with his bride. He made special arrangements for Elvira to ride in state: she had a special seat with a roof and was carried by two bearers. Elvira must have felt like a queen as she proceeded into the African unknown.

When Martin reached Kikuyuland he selected the flat spot that he thought best for his camp. It was near a water source – the Nairobi river was near by – and just below a hill (later to be known as Nairobi hill). When the tents were put up and all the basics completed, he left some of his askaris to guard the camp and continued on to Fort Smith to show Elvira the next station, at which they arrived on 20 September 1896.

Martin in true fashion arranged celebrations, inviting all his friends so that they could be introduced to his new bride. There was quite a gathering: James Wallace and his wife Mary came to Fort Smith, as did two other married couples, Henry and Helen Boedeker and James and Mary McQueen. Both Mrs Wallace and Mrs McQueen were pregnant.[10] Martin had also arranged for his bride to officially open the Ngong Fort that same month, so they travelled down, probably with the other couples. Elvira, dressed in her smart Victorian dress, hoisted the flag 'to the sound of two bugles frightfully out of tune'.[11] The fort was named Fort Elvira in her honour. Martin had made a great effort to make Elvira feel that she belonged upcountry.

In September 1896, an Edward Russell also went to 'Nairobi' to see 'Mr Martin and his new bride'.[12] This is the first mention of Nairobi as a place. East African maps prior to 1897 show two sites indicated as 'Martin's Camp', one of them being on the spot where the future Nairobi railway station was established and the future capital of Kenya.

It was a little later that Elvira became pregnant, and after long discussions they felt it would be better if she returned to Zanzibar to have her baby. A recent Maasai raid on a caravan in the Kedong valley had killed over 600 African porters;[13] this unrest must have added weight to the decision for her to return to Zanzibar in early 1897. James, meanwhile, had to get on with the work of improving the camp for the railways.

It seems a fitting coincidence that Martin, who

was the known expert in leading the way into Kenya and Uganda and had already done much to open them up, was entrusted with the development of the camp that was to become Nairobi. It is also interesting that it was not part of an existing African/Kikuyu village. Nairobi owed its early success to being the centre and headquarters of the railway, halfway between the coast and Uganda. It then developed into an all-tribe centre, not belonging to any traditional tribe, and people from all areas were tempted into the city to seek the peace and safety that they did not have at home. It became a melting pot for inter-tribal friendship and was a refuge for those who had escaped conflicts and sought the promise of opportunity. It grew rapidly and symbolized the new colonial era more then any other place. Although there exists a street named after James Martin in Kampala, Uganda, there is nothing in Nairobi to indicate the city's earliest origins.

The years of 1896–7 had been a rollercoaster for Martin's emotions. He had achieved so much and fulfilled his dream of getting married. It was hard for him when his wife went back to the coast and he had to carry on with his job. His friends Mrs Wallace and Mrs McQueen both gave birth to their children upcountry – only the second and third white babies to be born there. Henry Boedeker was a doctor, so it is quite probable that he assisted at the births. (The first white baby to have been born away from the coast was J.A. Stuart Watt, at Negelani on 23 August 1895.)

Elvira gave birth in Zanzibar to her first child, a daughter, whom they named Magdalene. Unfortunately, in spite of the efforts of Elvira's father, who applied all his medical knowledge, the child died only a short time later on 11 December 1897. The family were distraught.

Opening Africa

Having been elated at the news of his Magdalene's birth, Martin was devastated by the sad news of the baby's death. He had up to that point seemed to live a charmed life in Africa, not succumbing to the many hazards and diseases that most white people suffered. The death of his daughter would have been a terrible blow to him. When Elvira returned they settled down to life together in Eldama Ravine. They were to have a second daughter three years later. She was born in Eldama Ravine Station and given the name of Eldoma. James Martin had founded a family at last, against the odds.

*****

Soon after he resumed duty at Ravine, in August 1897, there were rumours of unrest among the Sudanese troops in Uganda. Contingents of men had been deployed by their superiors in expeditions to various parts of the country, resulting in long marches. For months the men had not seen their families and had been allowed little rest. The pay was miserable and the supply of clothing delayed. A couple of British officers, who did not know the askaris' language, were unpopular. When a large force of some 300 men started on another long expedition to Juba on the coast to delineate the border with Italian territory, they had hardly got beyond Ravine when the men refused to obey orders. They stopped, declined to proceed any further and returned to Ravine, where they were addressed by Jackson and MacDonald, but they refused to parley and started marching back on the long trek to Uganda, with the intention of joining up with other Sudanese askaris. They got as far as Lubas, where they were surrounded by MacDonald and loyal forces

Founding a Family and a City

made up of Baganda and Swahili troops.

It was here at Lubas, where Bishop Hannington had been murdered, that there were armed clashes and mutineers shot three British officers, while a fourth officer was killed in action. The mutinous askaris were, however, contained in their fortified enclosure, and pinned there for several months until one night they managed to sneak away. They were followed by loyal troops, and were finally beaten and dispersed. By February 1898 the mutiny was over.

Meanwhile, back at Ravine Martin was having a worrying time. There were still a small number of armed Sudanese troops under his charge at the Station. With his excellent spoken Swahili and Arabic he had won their confidence and loyalty to the extent that he had persuaded the military commander to let them retain their arms – the only post in East Africa to be given this privilege while the mutiny lasted. Martin wisely took the precaution of increasing the number of armed Zanzibari porters in the camp.

In November 1897, in the midst of this tense situation, alarming reports were brought to Martin that a large force of over a thousand armed men with numerous camels was approaching Ravine from the direction of Abyssinia. Martin had not been told of the scientific expedition of Lord Delamere that had started at Berbera on the Red Sea and was proceeding through Somaliland and Abyssinia with the aim of reaching East Africa. He called his clerk de Silva and dictated a letter to be despatched immediately to the intruders as they reached Lake Baringo. It was not addressed to anyone in particular and read: 'Sir, Please take notice that you are now on British soil. Any act of aggression on your part will be sternly resisted!'

The barely legible signature looked something like 'J. Martin', and when Delamere asked the locals who the officer in charge of the area might be, he was told 'Bwana Martini', which made him wonder why an Italian was trying to control entry of his party into British East Africa. In the event, his expedition turned out to be much smaller and in no way threatening, but arriving unannounced as it did at the height of the Sudanese army mutiny, it put fear into Martin that his small district headquarters might have to oppose a large Abyssinian force advancing in aid of the Sudanese rebels. In fact he had sent a hurried message to warn the senior officer at Fort Ternan that armed help might soon be needed, and had immediately set his own men and porters to dig a deep trench around the station and to reinforce the protective boundary palisade.

On arrival at Eldama station, Delamere was welcomed by the 'wiry little man' who was Her Majesty's Collector of the District of Baringo and was entertained in the warm and lavish manner for which Martin was gaining a reputation. Delamere spent the night at Eldama Station and received an account of happenings in Uganda and in the outer world.

When the Sudanese mutiny was over early in 1898, Martin returned to his normal duties, which included those of 'Collector of Baringo District' – that is, he was responsible for collecting poll tax. He had the unenviable task of travelling on foot to tribal areas to collect a tax from untamed inhabitants who had never been faced with a demand for the payment of any sort of tax before. When he first visited the wild Suk tribe for this purpose, he had wondered whether he would come out alive as he was rapidly surrounded by fierce chanting warriors. Once again the resourceful Martin extricated himself by a ruse. Seeing some good Somali ponies that belonged to the chief, he suggested that the chief should exchange some of these ponies for some of Martin's powerful 'dawa' – consisting of a couple of bottles of potent brandy, which the Chief was not averse to tasting during the bartering process. Martin himself was a teetotaller, but this did not prevent the bartering being protracted until well into the night. The next morning the chief, feeling quite miserable and heavy in the head, pleaded for some more 'medicine' until eventually he was ready to agree to anything, including allowing some of his best ponies to be led away by Martin in payment of poll tax. Martin was shrewd, and had started trading on his own account; there is little doubt that only part of the haul of ponies was needed to cover the tax.

While he was employed with the British East Africa Company, practically all his time had been spent on the march and he had not had much opportunity for private trading. Admittedly he did not entirely neglect the chance of purchasing goods at a cheap price at the coast and selling them en route to Buganda or disposing of them at a profit in Kampala itself. There was plenty of ivory to be had in the form of elephant tusks, and the profit margin on its sale at Mombasa was good. It would not have been beyond the resources of the leader of a caravan returning empty to the coast to utilize the porters for this trade.

As a District Officer at Eldama Ravine, Martin was not supposed to engage in private trading, but there were no regulations stipulating this in the early days of the Uganda Administration. The first officers to be appointed were not formally recruited into the Service but were merely transferred en bloc from the IBEA Co. Through the staging post of Eldama passed all the private traders with their caravans. These were organized by Arabs striving to replace the profits of the slave trade by less questionable operations. There were also a few European adventurers who came to seek a fortune in East Africa, chief among them the red-headed Irishman Charles Stokes. He sold guns wherever he could and bought vast quantities of ivory with the proceeds. He combined his trading with the organization of large caravans to escort missionaries and carry supplies. Initially a missionary, he had married one or more Africans who bore him children. He dressed up as an Arab trader, though his pale skin and his thick red beard showed that he hailed from a different part of the world. He was reckless, broke all the rules and became rich.

During the six years that Martin lived at Eldama Station he added to his salary substantially by trading and was able to live in lavish style. He spent a good deal on entertaining guests and showering them with hospitality, and must have become the envy of less enterprising officers. His trading activities eventually attracted the attention of his superiors, and would have drawn serious censure on him had he not already been something of a legend in East Africa, and had he not had many friends and protectors, among them Sir Frederick Jackson. However, at a convenient moment Martin was eased out of his lucrative posting and sent to one of the islands in Lake Victoria with the task of building an official station there.

12

Benefactor in Entebbe (1904–1914)

Martin remained in the Ssese Islands for only a brief period, and was glad when he was posted to Entebbe as District Officer and Collector. He did not fancy the crocodiles or the hippopotami that abounded around the island shores, and he intensely disliked the mosquitoes and flies that intruded everywhere. In particular he was nauseated by the periodical appearances of the all-pervading lake fly. The lake flies bred in their millions in the waters of the lake, and rose like minute bubbles to burst on the surface and release an immense cloud of flying insects that filled and darkened the sky for miles around. They were impossible to avoid. The flies that were blown ashore by the breeze dropped down to earth and, dying, emitted a repulsive fishy smell that pervaded the atmosphere and hung around for days.

Entebbe was the rapidly growing centre of the British protecting power and had become the gateway to Uganda. Travellers arrived by the newly completed railway that steamed its way from Mombasa on the coast to its terminal point at Port Florence on Victoria Nyanza; thence they crossed by lake steamer to Port Alice at Entebbe. The long, laborious haul of goods and supplies for Uganda on the heads of Zanzibar porters had all but died out; the railway took over seemingly overnight, bringing commerce and progress and contact with the outside world all along the safari route.

For Martin, Entebbe was a mixed blessing. He had plenty to do as a District Officer in dealing with people's problems in and around Entebbe. He knew the local Baganda people and their Luganda language from some of his earlier trips, so much was familiar to him. He had arrived at a time when the future township was being laid out with roads and parks and new buildings – mainly offices and dwellings for government officials and for Indian and Goan workmen. The laying out of the Botanical Gardens on the lake shore, for instance, involved the engagement of over a thousand labourers. His official duties included meeting important visitors on their arrival and attending their departure at Port Alice pier.

Among these dignitaries were the Duke of Abruzzi who, with a party of Italian Alpine troops, had come on a mountain-climbing expedition to the Ruwenzoris – the fabled Mountains of the Moon. Martin accompanied him from Entebbe with an escort of twenty-six askaris and sixty-seven porters, and travelled with him, flying the British and Italian flags, to the foothills of the impressive mountain range. As a keepsake Martin was presented with one of two hunting rifles bearing the Duke's arms that had been specially made for the Duke by Greener. The gun is still in the possession of Martin's family.

Another visitor of repute whom Martin met at Entebbe was Winston Churchill. As Parliamentary Under-Secretary for the Colonies, Churchill arrived on 19 November 1907, clad in a white uniform and displaying an impressive row of medals. He was received by a guard of honour of Sudanese troops and was introduced to senior officers of the Protectorate. One of the first things that struck Churchill was the great heap of elephant tusks on the wharf awaiting shipment across the lake. There were no cars in Uganda in 1907 and Churchill travelled to Kampala by rickshaw – a distance of twenty-six miles, which was accomplished in three hours. The officers who accompanied him went on bicycles. After a short visit, Churchill left Uganda via Sudan.

The missionary order of the White Fathers found Martin very helpful. Their headquarters were at Rubaga, in Kampala, but they had set up a mission school and hospital at Kisubi, a couple of miles outside Entebbe. They now wished to build a church and adjoining house at Entebbe, and Martin assisted them in choosing a suitable site at Bugonga, an area of about two acres for which £25 was paid. He gave them a gift of the timber needed for constructing the benches. Later he presented the church with one of the three bells that now hang in the belfry. While the church was under construction Martin's house was used for religious worship, and he accommodated the father in charge until the church was ready. He also proposed that the mission should build a nuns' convent for the care and education of young Africans. He selected the site for the Entebbe cemetery, which he then divided into four sections: one each for the Catholics, the Protestants, the Moslems and the pagans.

Martin saw a good deal of the White Fathers at Entebbe, and he is frequently mentioned in the daily diary that they were obliged by the rules of their Order to maintain. There is a reference in the Mission diary to an inquiry conducted by Martin, on 20 July 1905 into the case of the sale of a girl from Kisubi by her uncle for thirty rupees. Her father had agreed to the deal, taking most of the money for himself. The girl was being forced to turn Moslem; she refused, escaped and sought refuge with the White Fathers. Around this time several young Baganda were sent by the Entebbe White Fathers to the island of Malta to be educated and trained as doctors. On their return, many of them made quite a name for themselves in government service during an epidemic of smallpox. Meeting them would have been the first contact Martin had in Africa with people who had lived in his country.

When James Martin finally left Entebbe, the White Fathers said he was a 'good and generous friend'. He does not appear to have been an unduly good practising Catholic, however. When, on the occasion of the marriage of his daughter to a Mr Tarrant at Entebbe, he attended Holy Mass, the Mission diary records that 'he departed from his usual habit in order to accompany Mr Tarrant at Mass'. Having said that, one must mention that his manner of life at sea and later on constant safari was hardly conducive to the practice of religion. After he was married in Zanzibar in civil form before the British Consul, he later took pains to have his marriage blessed by a Catholic priest at Eldama Ravine station, and he arranged for his daughter, Eldoma, to be christened.

Martin was now fifty years old. He was still very active and ready to put his hand to anything, but as the work of administering the country developed and became more technical and more decentralized, his position became more difficult. His usefulness lay in his spoken knowledge of several African languages, with Swahili and Luganda in the lead, as well as his resourcefulness and uncanny perception of what would impress the native mind. His popularity in Entebbe was unrivalled, while his long and vast experience of East Africa gave him a clear advantage over many other District Officers in the field. However, he came down badly in the matter of written instructions, reports or correspondence, as he was helpless without the aid of a trusted clerk or friend. The problems that arose because of his illiteracy had never been light and they now increased to the extent that he frequently became more of a burden than a help.

    The ingenious manner in which he covered up his lack of education caused much amusement. He avoided embarrassing situations by craftily skirting round a subject, or by breaking into one of his amusing anecdotes – for he was a past master of the art of recounting stories, whether genuine, fantastic or made up on the spur of the moment. His friends saw through this subterfuge, but Martin was so likeable that they accepted him as he was and made the necessary allowance for his weaknesses. He was renowned for his hospitality, and this no doubt helped. At the Entebbe club bar he was wont to offer rounds of drinks and good humouredly tell his friends: 'You sign – I pay!' On occasions he was known to avoid writing or signing his name by claiming that he did not have his spectacles with him – though none of the photographs of him ever showed him wearing spectacles. He is said to have repeatedly used this ruse when dealing with court cases as a junior magistrate. It seems impossible that anyone should have been allowed to conduct a trial with his degree of illiteracy, but his uncanny understanding of the Africans' mentality, of their customs and reactions to situations, made him a fair judge, even if not all of the technicalities of the law were always followed. Of course, he had to lean heavily on his clerk, and when the time came to read the sentence he invariably ask the clerk to read it.

Opening Africa

Another consideration that brought pressure on Martin's job was that once the British government had formed the East African Protectorate (having taken it over from the IBEA), they proceeded to introduce a more structured and formal government organization; the number of commissioners was thus increased. Originally there were fourteen (including Martin) but in 1904 this had increased to 240. The commissioners consisted of the old IBEA Co. employees, new recruits from other African countries and new recruits from England. In their quest for higher standards in the Secretariat and no doubt with the wish to 'move on' some of the older, more experienced administrators and replace them with freshly trained people, they relied on their bureaucracy. Martin was therefore subjected to a lot more competition, particularly for the good jobs, and one of their selection measures gave an indication of how they operated. Around 1907 the Secretariat brought in a new policy and all Secretariat officers had to undergo the following test: all the words of the following sentence had to be spelled correctly:

> It is agreeable to perceive the unparalleled embarrassment of the harassed peddler while gauging the symmetry of a peeled potato before a committee of judgement.[14]

This was a requirement for an administrator of the African territories. It seemed like a scheme concocted by the 'New Guard of Administrators' to clear out some of the older, less literate but more practical men. Whether the illiterate Martin was asked to take this test or not is not known; however, it was around this time that he moved on to another job. Martin, with all his experience and knowledge of the native languages, who had done much for the locals and supported the traders in difficult situations, must have felt the weight of injustice fall upon him. He saw the writing on the wall.

However, a move would, in all, be a welcome change for Martin, who had now been in the Secretariat for thirteen years. While it had given him a good position and great opportunities, he was not always at ease with the direction in which the Secretariat was going. Martin's own tendency to work with the locals and assist them with education, family and health would have been in conflict with the more 'military' stance of the Administration. He worked at the sharp edge and had to make decrees palatable and acceptable to the natives. This sense of conflict was shared by many of the missionaries.

Perhaps this definitive statement of the Administration's views, made by Sir Charles Eliot, the High Commissioner, in 1900, was a turning point. He said that it was only a few years since the place had been a human hunting ground, where the native tribes warred with each other to get slaves to sell to Arabs, and therefore:

> We are not destroying any old or interesting system, but simply introducing order into blank, uninteresting, brutal barbarism ... East Africa is not an ordinary Colony. It is practically an estate belonging to His Majesty's Government, on which an enormous outlay has been made and which ought to repay that outlay.[15]

Opening Africa

What Eliot suggested as a means to achieve this was an influx of white settlers to develop the country and utilize the railway. This would transform the whole protectorate. But Martin probably couldn't help thinking that this was not the way the British had administered Malta, and wondering whether it was the right way. He knew that the Africans valued their land, although it was clearly under-utilized, and taking it over, perhaps by dubious means, would have repercussions. This policy would clash with the fundamental motivation of the natives to 'own their piece of land'. Even areas that were uncultivated were still part of their land in their own eyes, and they had fought for it over generations. The Maasai, for instance, as semi-nomads, claimed the vast territories that they moved within to be theirs. There would probably be more conflict ahead.

Another thing that no doubt made Martin feel uncomfortable was the rigid hierarchies that the Secretariat had brought in. When he started there had been nineteen officers. The attrition rate was high: of the twenty people in post in 1887, seven had died by 1907 and the death rate of the newcomers was much the same. Mainly they died of disease – often blackwater fever, a malaria-like illness. Most of these people had arrived with no knowledge of the country. They were divided into higher officers and the lower officers, and each group had its own social club. They tended to see the rest of the new society they were forming as a hierarchy consisting of the whites, the white settlers, the new influx of Indians and the native Africans. Each of these groups had its own elaborate sub-hierarchies. The complexity of the African tribal system with its sixty or so tribes was made even more complicated with the addition of these hierarchies

Benefactor in Entebbe

– an intricate order to manage. As Martin's wife was a darker-coloured Portuguese she was sometimes taken to be Indian or Goan, which had led to one or two awkward racial situations: people sometimes assumed Martin to be in one social strata and his wife in another.

Although frowned upon by the Administration, a few bachelor officers in remote areas had native mistresses, or even concubines, and gave no status to their 'partners', which was again 'socially incongruent'. The resulting children of mixed race were ignored by the white community but they luckily found easy acceptance in the more tolerant African circles. The administrative climate exacerbated the racial divide rather then minimizing it, and it was a strange, changing and complex phenomenon. In any newly forming society, the initial vision or structure tends to perpetuate until an eventual crisis readjusts the social gulfs it has created. In this case the structures became more entrenched and more and more levels formed as time went on, rather than there being a movement into a single, more integrated melting pot. The new white administrators at the top spent less and less time talking to the grassroots African tribesmen in the way Martin had done.

*****

Early in 1907, an opportunity arose for Martin to be eased out of his job in government service without embarrassment. The manager of the large Mabira Forest Rubber Company was leaving Uganda on his retirement, and Martin was offered the manager's job at a higher remuneration than his own government salary. The solution undoubtedly accommodated Martin. He could live in comfort while controlling a large body of African labourers, something that was well within his competence. He could keep up his friendships in nearby Kampala and Entebbe, and his new home at once became known for its hospitality. When he visited his friends he went laden with gifts, and returned with presents that had been pressed upon him. But he did not know the value of money and made no provision for his old age.

His leisurely life at Mabira came to an end with the start of the Great War in 1914. Even at the age of sixty he insisted on doing his bit for the Empire by seeking a commission in the Intelligence Corps and going to the front at Kagera river, which marked the dividing line between the opposing German and British forces. The story is told of how Lieutenant J. Martin was pulled along the front line in a rickshaw wearing carpet slippers!

One fascinating anecdote is recorded about Martin's war service. The German light cruiser and armed raider *Koenigsberg* had been roaming the South Atlantic and the Indian Ocean sinking Allied shipping and spreading terror wherever she went. She even closed on Zanzibar Harbour when the British cruiser, HMS *Pegasus*, was having her boilers cleaned and was thus immobilized. The *Koenigsberg* sank HMS *Pegasus* at her moorings and then withdrew to German East Africa to escape the Allied warships that were now hounding her. She entered the winding Rufiji river and withdrew up it as far as she could, and there felt herself safe from naval attack, for the water was shallow and the refuge hidden from view. The *Koenigsberg*, however, could not escape her final destiny. The Royal Navy sent out from England a special ship with a shallow draught, a Monitor, armed with one sixteen-inch gun. The Monitor could navigate up the Rufiji close enough to hit the German raider. This she did, led on her way by a small tug piloted by James Martin.

13

Out of Africa

After the war it seemed that the world had changed. It was the second big change that Martin had seen: the first had been when the railway arrived in Nairobi. The war was won, but it seemed that the economy was the loser. The company he worked for had suffered with their rubber exports during the war and there was no work for him there now. The whole of East Africa, like many places in Europe, was in a depression, with little or no work anywhere. He was seen in Mombasa in poor circumstances, buying and selling chillies. His luck and his money had run out. The environment was harsh and he clearly could not cope in a new age that was so different from and so much more demanding than the one he had been used to. His family probably convinced him to return to Portugal where some of Elvira's relatives lived, and having got together the money for their passage they caught a ship bound for Lisbon.

Martin would have had time to stand at the stern of the ship and watch the churned sea wake created by the propellers, and look back at the outline of Africa as it receded into the distance. He must surely have pondered on what his thirty-six years there had all meant and perhaps tried to put the things he had seen and the people he had met and known into an honest critical perspective.

When he first arrived there sail had been faster

than steam, and moving a ship had involved setting the sails with every shift of wind and being totally dependent on the weather. Now there wasn't a sail in sight, and running a ship was no effort at all. Ships went faster, and in a straight line. Not having to use the unpredictable power of the wind, sailors were in much less danger of being shipwrecked. The change in travelling around inside East Africa was even more amazing: a three-month journey on foot to Uganda, with all its hardships and hazards, had been replaced by an easy few days sitting down in a train compartment.

It was, however, the lives of the people that had changed the most. When he had arrived in Zanzibar the signs of slavery were everywhere. It was by far the worst thing that was happening. It was a human tragedy for each one of the slaves captured and meant death for many of them. He had come to realize also the consequences it had for the lives and minds of the people who were not themselves taken as slaves. They lived in fear of the 'spirit men', and this dread and superstition was used to control their way of life. Martin understood this effect from his childhood days, although not in the same measure, but he had outgrown it and knew the importance of conquering false notions and their power.

It was also the cause of much inter-tribal conflict and it bred suspicion and mistrust. It was pure poison and had become ingrained in their cultures.

Slavery was a medium by which tribes or their chiefs could trade with the outside world and get essential products for their advancement, such as wire and beads for their work and personal decoration; food and grain in times of famine so that they could survive; and guns and weapons so that they could kill game to eat and also protect themselves against their enemy neighbours. It was difficult to quantify the massive change that removing slavery had brought. The creation of other forms of trade to enable the locals to buy what they needed without having to traffic slaves was an important factor in the change in which Martin had played a part. Developing honest trading bit by bit, the people slowly became more open to outsiders and trusted each other more. They began to work together; although tribal differences were always there, they became more manageable.

Martin had come to understand the natives' needs and problems, and that the vicious circle created by slavery had to be actively broken in order for progress to take effect. Knowing the cause of their anxieties and animosities often helped him to solve disputes and negotiate peaceful solutions, whether for the passage of one of his caravans or in the administration of his station at Eldama Ravine. He was gratified that he had potentially saved so many lives by being part of the group involved in ending the slave trade. It had been the cause of so much of the trouble and killing he had come across.

Regarding relations between the tribes, one thing that was clear was that whereas before each tribe was isolated and did not have much contact with others, now the Luo talked to the Kikuyu and the Luhya with the Nandi and so on. There was much more contact between the tribes, and not only in the melting pot of Nairobi. The Europeans had initially considered the indigenous people to be 'wild savages', but now the richness of their cultures and languages was known to many. The diversity and complexity of the different tribes that made up East Africa was a truly remarkable thing that no one had imagined at the beginning. He had certainly done his bit to proclaim the diversity of those societies.

The fact that Martin couldn't read or write was seen as a disadvantage by his white colleagues and it had held him back and caused him much frustration, but he found that the Africans acquired information in the manner that he did and that his own way was better for communicating with them. He stubbornly held onto his way of doing things; he believed that he had, in many ways, been ahead of his colleagues in communicating with the people because he had had to find a way to do it and he enjoyed it. He had been grateful to his friend de Silva who had helped him by doing his reading and writing. His illiteracy had kept him vulnerable, particularly in European social circles, he counteracted this with generosity, humour and know-how, thus maintaining the respect of his friends and colleagues. Consequently this humility had kept his feet on the ground. He must have regretted that his mass of knowledge about African people would not be written down and so would be lost.

Martin had, in the main, been altruistic in his approach to Africans. In terms of education and medical support and the prevention of disease, he had assisted the local people as far as he could, but this was not always easy. Some of the chiefs were autocratic and irrational in their behaviour and in judging their fellow men; the Kabaka (chief) of the Baganda in Uganda was an example. Martin felt that he should prevent this and condemn it at any local tribe democratic councils or judgements when he was acting as District Commissioner. He noticed that the position of women was low in tribal cultures, whether Kikuyu, Luo or Nandi. He had always worked to support respect for women. Whether by stopping female circumcision, or by reducing their work loads or marital disputes, he felt that he had helped many African women.

There was an image in his mind of the way he remembered the British to have acted towards his own people at home in Malta. They had seemed to be there to help the Maltese out and bring them trade and business and protect them from other powers. They honoured the existing institutions of the Maltese. There was no violence and no abuse of the Maltese, and no grabbing of land by incoming settlers. Yes, there had been the upper classes and the lower classes that clashed now and then, but the English had lived among and associated with the upper-class Maltese on an equal basis, it seemed to him. He understood that there was less of a gulf between the British and the Maltese, with them both being European, than there was between the British and the African tribes, but couldn't the same British approach that worked in Malta work in Africa too? He didn't often speak about this but it was what he felt.

The later white 'grab' of African land, no matter how much it was cloaked in contractual jargon, was not the best way. Disputes over property were always a problem. After the railways were built, many more white settlers were encouraged to move in and they bought up the natives' land, usually to local protest. The settler farmers who came from either South Africa or Britain were a varied group of people. Some built good farms and employed and taught farming to their native staff – they were fine, hard-working people who did a great job of clearing wild land for cultivation. They taught the local people that farming wasn't just about providing for yourself but was about working together to produce a surplus for others and for the country as a whole.

Out of Africa

There were others, however, who abused the blacks at any opportunity. The circumstances in which they were given or bought their land were dubious and often local chiefs were cheated into selling land without knowing it. Things became a lot clearer years later. Martin was glad he had never owned land; it had often been offered to him, but he wasn't a farmer and he instinctively felt that owning land was not the right thing to do.

There were also civil servants fresh out of Britain with their university degrees. They were willing and sincere but lacked knowledge of the local situation and very often tried to push views that were not always well received.

A category of people that he had met and didn't care for were the white ivory hunters. Their sole purpose was to kill as many of the magnificent elephants as possible and cut out the ivory: over 150 animals could be killed in a single trip. These people did have certain good points, however, and sometimes helped villages clear off elephants that were eating their crops. Once an elephant had been killed the vultures would circle above it, and this was the sign for the women of the village to go and hack meat from the dead animal, with the vultures pointing the way. The women would clamber into the carcass to get at the prized fat. An elephant kill was good meat and a celebration for the village, but the hunters' main purpose was greed and not any betterment of the locals. Some of them, like 'Karamoja' Walter Bell, had become very rich and established large estates back in their home countries, leaving little behind in Africa. This wild hunting still continued but had become more moderate, and the hunters' exploitation of slavery had stopped. This ivory trade had, however, played its part in developing business in Africa and Martin had himself dabbled in the 'white gold'.

The many missionaries that he had got to know did a good job in education, and that achievement should not be underestimated. Many people now had a better education than he had had as a boy, which was good progress. Some of those missionaries pushed religion too hard, and this seemed to him out of place in the tribal cultures. However, some of the tribal traditions were not good and the missionaries had helped to get rid of them: for example, the influence of the witch doctors generally had a negative effect and could be criminal. Some of the preachers lorded it over the natives without giving them a chance to develop their own culture. Some that he had met couldn't cope with the conditions and either went mad or became ill and simply couldn't take it any more. They paid an awful price for having wanted 'to come and help the natives'.

He admired his friends and colleagues at the IBEA and in the British colonial government, but again some of them had a Victorian 'British Empire' approach and often a military background that made them authoritarian and sometimes ruthless. They often took the military option – too readily at times – as Frank Hall had done at Fort Smith: he had learnt to deal forcibly with African dissent and then used the fear engendered to generate trust. (This was a typical blueprint for many of the colonizers.) Once when three of his mail men were murdered he undertook raids with 150 men and burnt and destroyed everything the rebels had. This was gross overreaction, and had given communication and negotiation little chance. Nevertheless, Frank went on to do much good work on road-building and became well-respected by all including the Kikuyu.

Martin probably realised that in many ways he had done things differently, and his far from perfect upbringing in Malta and luck had helped him survive, which was the exception.

There were other sad stories concerning some of the more aggressive natives – such as the Wakwafi tribe (a part of the Maasai) and their warriors – who after much conflict with caravans were eventually subdued by a series of disasters: rinderpest, a viral disease, came from the north (Red Sea) and killed their cattle, and in 1889 smallpox and famine depleted their numbers. Internal conflict, disease, poverty and hunger further reduced the tribe.

These were some of the regrettable aspects of what he had been a part of. It could all have been done better. However, when he thought of his old camp in Kikuyuland, which had grown into Nairobi in such a short time, he could only laugh to himself with satisfaction. Nairobi was a refuge for people of all the different tribes who needed to escape situations in their local villages. It offered hope, and a way of mixing tribal living and modern opportunity. He must have also recalled all his African friends and the chiefs and remembered the good times they had had.

His African friends did not forget him and seemed to have paid him the ultimate complement in naming their children after him. There is evidence of at least one Kenyan from the Kisumu / Luo region, born in 1975 who bears the name of James Martin.[16]

*****

Very little else is known about the life of our remarkable hero. No doubt in the few remaining years of his life he made new friends and kept them entranced with his stories.

James Martin, alias Antonio Martini, died in 1925 and was buried in Lisbon. He was sixty-seven years of age.

Martin in retirement:
sketch by Philo Pullicino

Epilogue 1: The author's discovery of Martin's story

Twenty-five years after Martin had severed all connections with East Africa, the author of this book – blissfully unaware that Martin had ever existed – embarked with his young family on SS *Almanzora* in the Grand Harbour of Valletta, bound for Zanzibar, to take up his first appointment as District Officer in His Majesty's Colonial Administrative Service.

Except in the higher echelons, officers once appointed in the Colonial Service were not readily posted to other territories, unless their transfer were hastened by evidence of either misbehaviour or outstanding merit. Having managed to avoid both of these counts, the author remained in Zanzibar for seven years until the Governor of Uganda, under whom he had served during the 1939–43 siege of Malta, arranged for his exchange with another administrative officer. The seven years in Zanzibar were followed by ten years in Uganda spent entirely in Entebbe and Kampala.

During his sojourn in Zanzibar he worked in the Secretariat, which was housed in the old Victorian building known as Beit el-Ajaib, and handled the personal and public affairs of His Highness the Sultan, as his Private Secretary. He travelled with the Sultan on several official and state visits to African countries, and to the United Kingdom for the Coronation of Her Majesty Queen Elizabeth II: the Sultan's fourth attendance at a Coronation in Westminster Abbey. The author's quarters in Zanzibar Town changed several times as he went on leave to Malta and returned. As suitable premises for a family of five were not easily come by, he found himself living in the large premises known as the International Anti-Slavery Bureau, and subsequently in the building called 'Mambo Msiige' (which means 'don't copy this!'), which in the nineteenth century had housed the British Agent and Consul General and also General Sir Lloyd Mathews, Chief Minister of the Sultan. Other quarters allocated to him for varying short periods included the former German Consulate and part of the French Mission building. His chief recreation was sailing in Zanzibar harbour, tacking around graceful Arab dhows in sail or at anchor, and skirting the buoy that marked the position of the sunken wreck of HMS *Pegasus*, which had been destroyed in 1914 by the German raiding cruiser *Koenigsberg*. Another popular sailing destination was Grave Island, where stood the memorial stones over the last resting place of the Royal Navy officers and ratings who died in Zanzibar waters while suppressing the seaborne slave trade in the nineteenth century.

From Zanzibar, the author went on visits to Taveta and Moshi in Chagga country, in Tanganyika – formerly German East Africa and now Tanzania – and gazed on the fertile expanses and rolling plains of the foothills of Mount Kilimanjaro. He often journeyed to Nairobi and to Entebbe to attend inter-territorial conferences, the trip taking no more than a couple of hours by air or two to three days by railway.

When in 1954 he left Zanzibar on being posted

to Uganda, he worked in Entebbe and Kampala. In Entebbe he lived in a bungalow at Bugonga village close to the little Catholic parish church. He took his family on walks to Old Entebbe to see the monument erected to mark the spot where the first White Fathers, Father Lourdel and Brother Amans, set foot in Uganda on 17 February 1879. Originally erected at the water's edge, the monument alternately moved inland or waded out into the lake, following the fall and rise of the level of Lake Victoria. The pier for the lake steamer at Port Alice was a favourite spot for fishing, or for watching hippos or enjoying the sight of the swift dives of cormorants, pelicans and brilliantly coloured kingfishers in their ceaseless quest for fish.

Three of the seven hills around Kampala were crowned by temples of the three religions whose followers had fought each other so bitterly at the end of the nineteenth century: the glistening white mosque on one side and the Anglican and Roman Catholic cathedrals across the city on their original sites at Namirembe and Rubaga. The Kabaka's palace enclosure was near by and all were within range of Lord Lugard's Fort, which had made such a massacre of the Africans during the religious war of the winter of 1892. There is not much more than the outline of the foundations of Lugard's Fort left now on Kampala Hill, and little to indicate the important role it played. A stone's throw from the fort there is a street bearing the name 'Martin Street'. When it was first spotted by the author the name conveyed nothing to him, but a query to a passing Muganda gentleman produced the unexpected answer that the street was named after a remarkable and much respected friend of Buganda, a Maltese who had lived in Uganda in the first part of the twentieth century and enjoyed universal popularity.

This was the author's introduction to the saga of James Martin. It was inevitable that he should investigate further the unknown background of this compatriot of his who had been privileged to have his name displayed in the most historic part of Kampala. There was nothing at the time to indicate that his researches would uncover a well-documented account of the life of an extraordinary person who might justly claim a small part in transition of East Africa from being unknown to being one of the most fascinating regions of the world.

As his research into the Martin story progressed, the author gradually became aware of an uncanny destiny that led him, many years later, to live in the same places, cover the same ground and work in the job that formed an intimate part of James Martin's story.

Philo Pullicino 1994

OBITUARY of PHILO PULLICINO
The Sunday Times of Malta
20 August 1995

Philo Pullicino: a tribute
Dr Richard Manche MD

It is seldom that one grieves for a friend twice but this has happened to me, Philo being the cause of my grief. I knew Philo since early boyhood and admired his charisma from afar when he was superintendent of the Special Constabulary, and later when he joined the Colonial Service as a Secretary to the Governor of Bermuda. I saw him again, perhaps not really by chance, when he lured me to Uganda with promises of a beautiful country, the Garden of Africa, and a delightful happy people who were desperately in need of doctors. I followed him there and found more than he had described, especially in the medical field where I discovered that a paradise of scientific research of medical experiences awaited me.

Philo was there my host and mentor in his characteristic way – ever happy and, more importantly, devoutly Catholic. Through him I was introduced to the White Fathers Missionary Society and the other Catholic and Protestant religious orders tilling the field of God in the country that produced the Uganda Martyrs, canonized by Paul VI in 1969.

Philo and his beautiful wife Laura were the leaders of the Catholic communities of Entebbe and later Kampala, whence they moved when Philo became Clerk of the Parliament Assembly in 1960. Here his strong Catholic principles and his upright character, his devotion to duty and his meticulous attention to detail, coupled with his legal brain inherited from his ancestors, made him the idol of the Ugandan parliamentarians who sang his praises and showered honours on him on his retirement.

This was my first grieving for Philo, as when he left Kampala I felt I had lost a father, a guide, a philosopher and friend. I stayed on in Uganda until Amin made my stay uncomfortable, but Philo had gone on to higher things and had been appointed ambassador for Malta in Rome, representing his beloved island home in the Council of Europe and other European states. Here the government could not have made a better choice. This gentleman among gentlemen, with his aristocratic bearing and his vast western culture, did more for the Maltese image in Europe than anybody who came after him.

On his return to Malta he immediately hurled himself into voluntary works and public duties, culminating in membership of the Public Services Commission, where his long association with the civil service in many lands came in very useful to the country. Among his favourite voluntary organizations was the St John Ambulance, where he distinguished himself on the executive council both here and in Uganda. For his contributions he was knighted by the Venerable Order. He was also honoured by the Sultan of Zanzibar, whom he had served as private secretary in the fifties, and by the Queen for his achievements in Uganda.

His involvement in the erection of the Siege Bell memorial was perhaps his last contribution to the national heritage. His garden and his bridge club kept him smiling in his later years. It was my misfortune to be with him one evening when he heard the call to higher duties, and two days later, after my colleagues at St Luke's hospital had tried desperately to save his life, I was there to see him slip away on the long voyage home, leaving me grieving for the second time for my friend.

Six months have passed since his death. I have not seen his elegy in print. May this be a humble tribute to a great Maltese gentleman. May he rest in peace.

R. MANCHE
St Julian's, Malta

Epilogue 2: A token of tolerance

I took up my father's manuscript as it was not published before he died. I read the latest relevant publications on East Africa, and by using the internet have been able to add some aspects to this account of James Martin's life. The story fascinated me as much as it did my father, and one of my 'Eureka!' moments was discovering additional proof that Martin had 'founded' Nairobi.

As the story reflected things that my father himself had experienced in his life in East Africa, so too did it awaken my own memories. I was born in Malta and had lived in Zanzibar from an early age and went on to boarding school in Nairobi: my formative years were also spent partly in the wilderness north of Thomson's Falls, Entebbe and Kampala. I only left to move to university in the United Kingdom. I spoke Swahili, as we all did, and developed a love for the local people, their humour and customs, and the beautiful country.

Many years after I had left East Africa I went to the Tanzanian Embassy in Riyadh to obtain a visa to visit Zanzibar. The African ambassador, seeing my passport, called me into his office.

'Did your father ever work in Zanzibar?' he asked.

'Yes,' I replied.

'Well, it is your father I have to thank that I am sitting here to day. He arranged for me to get a scholarship to study in the UK. Please give him my kindest regards.'

I later reflected on the developmental journey

he and his ancestors must have made. It was these little things that told me my father had made a difference in East Africa. I was often lulled into the belief that the British colonial powers and their administration had exploited and ruined the countries of Africa. Yes, there had been exploitation, inappropriate forms of government, wrong delineation of territories – and looking at some of the African governments today, and the poverty that their countries are in, one might well wonder what good the colonial era did to these East African countries. But although there were greedy, selfish and ruthless colonialists who took from and abused the country and its people, there were also those who worked for the genuine betterment of the inhabitants, with no gain to themselves. The missionaries with their schools were at the forefront of this good altruistic work. Good wheat was sown among the weeds.

The scars of the slave trade and the resulting inter-tribal feuding could have only been patched up by some form of binding, overruling administration. Keeping each of fifty tribes as separate countries would not have been a solution for peace and prosperity. The railways were used to bring people closer together, to break down the isolation of the tribes and to increase the effectiveness of their honest trade, as well as to serve the security and economic goals of the government.

Looking back, the reality, in fact, is that this good work and process did set indigenous people free. The process may have led to the development of inadequate governments today, but it did allow the people to move on from being local hunter-gatherer tribesmen, in the late eighteen hundreds – with the poison of slavery eating into their society, tribes and villages and keeping them

A token of tolerance

captive with indigenous violence. The awakening of the land, bit by bit, by various outsiders, set the process of freedom on its course.

As a child at school – remembering the clop, clop of the askaris' boots walking up and down the corridor outside our dormitory with sub-machine guns, all night long – I didn't understand what the Mau Mau independence fight was all about, or what it meant when a fellow pupil (Robin Tui) left school on Friday not to return on Monday, hacked down in a forest near his home. Martin's story has now, after all these years, given me a better understanding of what it was all about. The reversion of the ownership of the land to the indigenous people who originally owned it, and their gaining their vote in a true democracy, only came after an uprising and fight led by Jomo Kenyatta that claimed many lives in Kenya. The government was at that time not close enough to the people to understand what was happening. They were unable to effect change through peaceful means and even broke their own laws in putting down the revolt. Perhaps there was not enough of a 'Martin approach' to the people.

The colonial administration made many mistakes – there is absolutely no denying that – but it did give countries the structures and institutions that enabled them to run their own democracies and bridge the tribal animosities. African countries that, like Sudan, did not benefit from similar structures appear to be worse off today. Failures have been due to the new African leaders rather than to the systems they sometimes blame: the good leaders are proof that the basis was a useable platform upon which to begin to work. Individuals, rather than governments, have been able to fulfil their

Opening Africa

potential as being among the most respected leaders in our world today, 130 years later.

Barack Obama is an example of this. Obama's grandfather, Hussein Onyango Obama, was born a year after the first white men visited his tribe; that was in 1884, the year in which Joseph Thomson and James Martin passed through his neighbourhood near Kisumu on the shores of Lake Victoria (see Chapter 6). They were probably the first white people to visit Obama's area. Martin was also probably the first white person to speak the local Luo language, and he initiated the interface with the Luo people in the 1900s. So Obama's grandfather was of the first Luo generation to grow up interacting with the 'European world'.

With the advent of the first white men, other traders followed and missionaries arrived to set up schools. Obama's grandfather would probably not have had much schooling in the early years of setting up schools and he went on to work as a domestic cook. It was Obama's father rather than his grandfather who benefited from this first educational offering. His father, after completing his schooling, was given a scholarship to the USA, where he met and married an American woman. The brief marriage resulted in his son, Barack. His father then separated from his wife and went on to be an economist in the Kenyan Government. Barack carried on his education with his mother in the USA (Hawaii). The story of Obama's links to the Luo tribe in Kenya is told by Barack Obama himself in his book *Dreams from My Father*. James Martin's story fills in some of the gaps to explain how the Luo tribe came to be where they were and how their education developed. It briefly covers what happened before Obama's father was born, and how the

A token of tolerance

meeting of the cultures – African and European – came about. There are many similar stories in East Africa, but perhaps none as spectacular.

It is a sobering thought that the opening up of the Luo region led, among other things, to the development of a man like Barack Obama, within only two generations. This could be seen as a result beyond the wildest dreams of the missionaries who started out there in the early days. James Martin, and the many like him – and indeed those of all races who are carrying on the work there today – will eventually get recognition for all they have done to open up East Africa. The work is still far from finished. When the expansive development of a man such as Obama is mirrored across Africa, what a massive future it will be.

Much of Martin's story concerns the development of tolerance between cultures. Several journeys of social change cross paths in this biography. The first is the journey of progress of the black Kenyan, as exemplified in Barack Obama and described above: it covers the 120 years from Thomson's and Martin's first meeting with the Luo tribe to Obama's US nomination. Another is the 'social journey' of the people of East Africa, and of Africa in general, and their governments. This journey has so far taken several generations.

James Martin considered Africa to be his home, but he knew he did not belong there. It did not feel like his country and when he retired he returned to Europe. A similar realization has probably been made by the multitude of non-black missionaries, doctors, administrators, teachers, farmers, engineers and tradesmen who have made their living and home in East Africa over the past hundred years or so.

Opening Africa

The social journey towards acceptance and respect, both between tribes and towards outsiders, is still only at the halfway stage.

M.J.P. July 2008

Appendix 1: Tribes and languages of East Africa

Uganda Tribes

Bantu-speaking tribes
Central Region: Baganda
Western Region: Batooro, Banyoro, Bakiga, Bafumbira, Bakonjo, Bamba, Banyarwanda and Batwa
Eastern Region: Basoga, Banyuli, Bakenye, Bagishu, Bagwe,
North-Eastern Region: Bateso, Jopadhola and Karimojong, Kumam, Jonam, Sebi, Pokot (Suk) and Tepeth

Nilotic-speaking tribes
North Region: Acholi, Alur, Langi, Lugbara, Madi, Kakwa
North-Western Region: Lendus (also in Zaire)

Kenya Tribes

The hunter-gatherers
Boni, Dahalo, El-Molo, Ndorobo and Sanye

Bantu-speaking tribes (agriculturists)
Western Region: Luhya, Kisii, Kuria, Gusii
Central Region: Kikuyu, Kamba, Meru, Embu, Tharaka, Mbere
Coastal Region: The Mijikenda (Digo, Duruma, Rabai,

Ribe, Kamba, Jibana, Chonyi, Giriama and Kauma); Segeju, Taveta, Pokomo and Taita

Nilotes and Paranilotes (pastoralists)
Nilote: Luo
Teso: Iteso, Turkana
Maasai: Maasai, Samburu, Njemps, Chagga
Kalenjin: Nandi, Kipsigis, Elgeyo, Sabaot, Marakwet, Tugen, Terik, Pokot

Cushites (shepherds)
Somali, Rendille, Galla, Borana, Gabra, Orma, Sakuye

Swahili (fishermen)
Bajun, Pate, Mvita, Vimba, Ozi, Fundi, Siyu, Shela, Amu

(Adapted from Nicholls, *Red Strangers*)

Appendix 2: Journeys on foot in East Africa performed by James Martin

1883	February 1883 to May 1884: sixteen-month safari with Joseph Thomson 'Through Maasai Land'.
1885	16 June: Journey to Chagga country with General Matthews. 11 September to 5 November: safari up Tana river with Sir R.G. Harvey.
1886	With General Mathews on military expedition to Chagga country.
1886	November 1886 to May 1887: leader of big-game hunting safari of Captain John Willoughby to borders of Maasailand.
1888	February to November: leader of safari of Sir Robert Harvey.
1889	18 June to 14 April 1890: leader of caravan of Frederick Jackson and Ernest Gedge from Mombasa to Uganda..
1890	14 May to 4 September: return to Mombasa in caravan with Jackson.

1890	Caravan to Uganda with Jackson leaves Mombasa 8 December; arrives Kampala 31 March 1890.
1891	7 April to August: return safari to Mombasa.
1891	September to 9 December: safari caravan to Uganda.
1892	8 January to 11 May: return safari caravan to the coast.
1892	August to 6 December: caravan leaves Mombasa for Uganda.
1893	January to May: return journey to Mombasa with large cargo of ivory.
1893	31 July to October: fresh caravan for Uganda leaves Mombasa with Captain Eric Smith.
1894	Return journey to coast.
1894	8 September: Uganda bound with W. Hobley, Foaker and Macallister.
1894	Martin joins Uganda Government Administration and goes to his new post at Eldama Ravine.

Appendix 2

1895	Martin goes to Mombasa and then to Zanzibar to get married.
1896	Martin returns to Ravine Station.

The above is not a comprehensive list, but every safari mentioned has been recorded in one or more books dealing with the early days in East Africa. The Royal Geographical Society stated that Martin had made no fewer than twenty-three trips travelling from Mombasa to Uganda or back. The distance from Kampala to the coast was estimated to be 800 miles.

Appendix 3: Martin's age

According to the Uganda Pension List in Uganda Blue Book he was aged sixty-nine in the year 1923: that is, he was born in 1854.

According to Sir Frederick Jackson, in his *Uganda Early Days*, he was born in 1857.

According to the official marriage certificate issued by the British Consulate in Zanzibar in 1896, he was born in 1860.

No trace of his Certificate of Birth has been found in Malta although there is a reference to his birth in il Marsa, Malta, in 1857.

Notes

1. Jeal, *Stanley*. p. 35
2. Jeal, *Stanley*, p. 15.
3. Kapuscinski, *The Shadow of the Sun*, p. 82.
4. Alpers, *East African Slave Trade*.
5. Kapuscinski, *The Shadow of the Sun*.
6. Jeal, *Stanley*, p. 4.
7. From a passage in Nicholls, *Red Strangers*, p.5; also with reference to Google Earth.
8. Nicholls, *Red Strangers*, p. 6.
9. Nicholls, *Red Strangers*, p. 18.
10. Nicholls, *Red Strangers*, p. 33.
11. Nicholls, *Red Strangers*, p. 23.
12. Nicholls, *Red Strangers*, p. 23.
13. Nicholls, *Red Strangers*, p. 38.
14. Nicholls, *Red Strangers*, p. 76.
15. Nicholls, *Red Strangers*, p. 47.
16. Partridge, *Unsung Hero*, p. 111

Bibliography

Alpers, Edward A., *East African Slave Trade*, Nairobi, 1967.
Bell, Hesketh, *Glimpses of a Governor's Life*, Sampson Low, Marston, London, 1946.
Blake, George, *B.I. Centenary 1856–1956*, George Blake/Collins, London, 1956.
Cauchi, N., *Maltese Migration in Australia*, Malta, 1990.
Cooke, Sir Albert, *Uganda Memories 1897–1940*, The Uganda Society, Kampala, 1945.
Dawson, E.C., *Bishop Hannington: First Bishop of Eastern Equatorial Africa*, Seeley & Co, 1887.
Filippi, Filippo de, *Le Ruwenzori*, French translation by A. Poizat.
Gray, Sir John, 'Mutesa in Buganda', *Uganda Journal*, Vol. I (January), 1934.
Hill, M.F., *Permanent Way: The Story of the Kenya & Uganda Railway*, E.A.R.& H., Nairobi, 1949.
Hobley, C.W., *Kenya: From Chartered Company to Crown Colony*, H.F.& G., Witherby, London, 1929.
Horden, Charles, *Military Operations. East Africa. History of the Great War*, HMSO, London, 1949.
Hunter, J.A., & Dan Mannix, *African Bush Adventures*, Hamish Hamilton, London, 1954.
Huxley, Elspeth, *White Man's Country*, Macmillan & Co., 1935.

Ingrams, W.H. *Zanzibar – Its History and Its Peoples*, H.F & G. Whitherby, London, 1931.

Jackson, Sir Frederick, *Early Days in East Africa*, Edward Arnold & Co., London, 1930.

Jeal, Tim, *Stanley: The Impossible Life of Africa's Greatest Explorer*, Faber and Faber, 2007.

Kapuscinski, Ryszard, *The Shadow of the Sun, My African Life*, Penguin Books, London, 1998.

Lyne, Robert Nunez, *An Apostle of Empire: Life of Sir Lloyd, William Mathews*, Allen Unwin Ltd, London, 1936.

Macdonald, J.A.L., *Soldiering and Surveying in British East Africa*, Edward Arnold, London, 1897.

Mason, A.T., *Nandi Resistance to British Rule 1890–1906*, E.A. Publishing House, Nairobi, 1972.

Miller, Charles, *The Lunatic Express*, Macdonald & Co., London, 1972.

Nicholls, C.S., *Red Strangers: The White Tribe of Kenya*, Timewell Press Ltd, 2006.

Obama, Barack, *Dreams from My Father*, Canongate, 2007.

Pakenham, Thomas, *The Scramble for Africa 1876–1912*, Wiedenfeld & Nicolson, London, 1992.

Partridge H.J, *The Unsung Hero*, PEG Ltd, 2002.

Patterson, J.P., *The Man Eaters of Tsavo*, London, 1934.

Perham, Margery, *Lugard: The Years of Adventure 1858–1898*, Collins, 1956.

Rotberg, Robert, *Joseph Thomson and the Exploration of Africa*, Chatto & Windus, London, 1971.

Royal Geographical Society, *Geographic Journal*, Vol. LXVI, No.1 (July, 1925).

Smith, Mackenzie & Co. Ltd, *History of W.Boyd & Co.*, W. Boyd & Co. Ltd., Nairobi,1949.

Snoxall, R. A., 'James Martin (Antonio Martini), 1857–1924', *Uganda Journal*, Vol. 1, No. 2.

Thomas & Scott, *Uganda*, Oxford University Press, 1935.

Thomson, Joseph, *Through Masailand*, Sampson Law, Marston, Searle & Rivington, London 1893.

Vandeleur, Seymour, *Campaigning in the Upper Nile and Niger*, Methuen & Co., London, 1898.

Vasco, André, *Zanzibar sans les esclaves: La fin du Ramadhan a Zanzibar*, Leopoldville, 1954.

Willoughby, Sir John, *East Africa and its Big Game*, Longmans, Green & Co., London, 1889.

In addition, the following documents and correspondence were consulted:

Uganda Journal, Vol. I, No.2, Staff List for 1895.

Uganda Journal, Vol. XXIII (reprinted at InvictaPress) Headley Brothers Ltd, 189 Kingsway, London), Captain Eric Smith's Expedition to Lake Victoria in 1891' by H.B. Thomas.

Uganda Journal, Vol. XXIII, No. 2, 'George Wilson and Dagoretti Fort' by H.B. Thomas.

Kenya Weekly News, Nakuru, 20 and 27 September 1957, 'Little Martin' by Rosa Walker.

East African Railways & Harbours Magazine, Vol. II, No. 9 (June 1956), pp. 291–3, 'The Development of Nakuru' by Mervyn Hill.

Zanzibar Gazette, 18 May 1892.

Despatch from Sir John Kirk to the Earl of Roseberry, 5 June 1886.

Bibliography

White Fathers: Diaries, Letters and Reports on the Missions of the Province of the White Fathers (1891 –1894), General House of the White Fathers, Rome.

Letters and research papers exchanged between Eldoma Winkler (daughter of James Martin), Dr Louis Galea, Sir John Gray, H.B. Thomas and Philip Pullicino.

The Author

Philo Pullicino was a man of many talents.

After leaving St Aloysius Jesuit College and University he joined the Malta Civil Service. At the outbreak of the Second World War he established the Special Constabulary which he commanded.

In 1943 he was seconded to British Honduras for two years as Private Secretary to the Governor. This was followed in 1947 by a posting to Zanzibar, where among his other administrative duties he was Private Secretary to the Sultan.

In 1954 he was transferred to Uganda at the request of the Governor. Here he served as an Administrative Officer in the Ministry of Health until Independence when he was appointed Clerk to the Uganda parliament, a post he occupied with great panache and expertise, both of which were greatly appreciated by the Kabaka and the Uganda Government.

Back in Europe, in 1965 he was appointed Ambassador to the Holy See and the Council of Europe and later to Italy and Austria, Switzerland, Greece and Israel.

He retired in 1972, but he served on the Public Service Commission from 1988 to 1991, until he decided that he should hangup his boots.

However, that was only the beginning of another side to Philo. He started painting in oils and sculpting in stone and was a dab hand at making miniature

The Author

'Maltese' clocks including a touch of gilding. He was an enthusiastic gardener and kept vines and fruit trees and played a good hand at bridge.

He mastered the computer and started his writing career which includes this book about the times and adventures of another Maltese who left a mark on East Africa, 'Jimmy' Martin.

To those of us who know East Africa well and Uganda in particular, the many details are a very welcome reminder of the facts, people and the place that was our home for many years.

Opening Africa

Standing: Mr de Silva, Charles Kitchen, Francis Dugmore; *middle:* Charles Lane, Elvira Martin, Frank Hall; *front: James* Martin, 1897

Standing: Frank Hall, Fredrick Jackson, Lennox Berkeley, Eric Smith, Fredrick Snowden; *seated:* Mrs Snowden, Helen Boedeker, Charles Lane, 1897

Fort Smith, 1890

Nairobi, 1899

Opening of the Railway, 1899

Nairobi, 1904

An African Family at Njoro, 1905

Zanzibar, Beit el-Ajaib, built in 1883

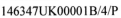

Lightning Source UK Ltd.
Milton Keynes UK
17 November 2009

146347UK00001B/4/P